LIFE ASCENDING

Alexander F. Skutch

 # Life Ascending

UNIVERSITY OF TEXAS PRESS, AUSTIN

Requests for permission to reproduce material
from this work should be sent to Permissions,
University of Texas Press, Box 7819,
Austin, Texas 78713.

LIBRARY OF CONGRESS CATALOGING IN PUBLICATION
DATA
Skutch, Alexander Frank, 1904–
    Life ascending.
    Includes index.
    1. Biosphere—Philosophy. 2. Biology—
Philosophy. 3. Evolution—Philosophy. 4. Human
evolution—Philosophy. I. Title.
QH 331.S53 1985   574'.01   84-21879
ISBN 0-292-70374-0
ISBN 0-292-74644-X (pbk.)

To Derek Goodwin,
whose books I enjoy

Contents

Introduction: A Planet at the Crossroads

 A few million years ago, before the last glacial epoch, Earth was no less beautiful than it is today. Many regions that in recent centuries have been devastated and polluted by man must have been far more beautiful. In that distant age, the sky was doubtless as blue, clouds as white, landscapes as green, tropical seas as ultramarine, mountain streams as sparkling as we have ever seen them. Trees grew as tall and stately, flowers were as lovely and fragrant, birds as abundant and colorful and songful, butterflies as beautiful as they are today. Mammals great and small leapt through the woodlands, grazed on the flowery prairies, burrowed underground, or swam in the seas. Oceans teemed with multitudes of fishes, cephalopods, crustaceans, and many lesser creatures. The planet was filled to capacity with life in myriad forms. And doubtless, to the measure of their psychic development, these creatures enjoyed their existence. Life was precious to them, and they clung to it with all their powers.

What, then, was lacking on Earth before man appeared? What could continued evolution accomplish, beyond keeping the living world adapted to ever changing climates and habitats and playing fresh variations on old themes? What could time bring forth that it had not already achieved during the billions of years since life arose on our planet?

Although animals enjoyed their own lives, did they appreciate the beauty of the living things around them, of the verdant land that bore them? Did they stand beneath the stars in wonder and speculate about the magnitude of the cosmos? Did they ask how they came to be here and what destiny awaited them as individuals and species? Did they study the ways of the creatures that shared Earth with them and sympathize with their joys and pains? Were they capable of compassion or mercy? Above all, were they thankful for the privilege of living on so fair a planet?

Aesthetic delight, wonder, scientific curiosity, sympathy, compassion, grateful appreciation of the boon of life on a beautiful planet—while perhaps not absolutely lacking, these psychic attributes appear to have been, at best, in a rudimentary state before man arose; even today they are poorly developed in much of mankind. To bring forth and perfect them was the difficult task, costly in time and pain, that awaited evolution after it had covered the planet with life in an endless variety of beautiful or bizarre forms, as it had already done many millions of years ago.

To create beings able to understand and appreciate, capable of gratitude as well as enjoyment, appears to be the end and goal of the world process. What would be the worth of a Universe spread over billions of light-years of space, if not indeed infinite numbers of light-years, containing millions of galaxies and trillions of stars and an unimaginable number of planets, yet devoid of consciousness? A Universe with no single being to enjoy existence in it, celebrate its beauty, or wonder at its immensity would be a barren Universe. The cosmos achieves value and a reason for existence exactly to the measure that it contains beings who enjoy, appreciate, and understand it.

Apparently, this is what, from the beginning, the Universe has been striving to achieve. We sometimes speak of an immanent or unconscious purpose; but this is a perplexing concept, and we might simply say that the Universe was so constituted or set that, given enough time, it would not fail to bring forth beauty along with beings able to enjoy and appreciate it. We cannot even imagine how the Universe came to be set to develop as it has done, yet the nature of the setting is fairly obvious. Atoms are social beings with a strong tendency to unite in patterns of ever increasing amplitude, coherence, and complexity—the process of harmonization. In favorable conditions, such as are found only in extremely restricted parts of space, the increasing complexity of atomic and molecular patterns begets living substance which, continuing the movement that produced it, assumes the most diverse forms and becomes capable of the most varied functions. Man, a late product of this aeonian movement, far from being a cosmic accident, is in certain aspects a partial fulfillment of the world process. He brings to his planet qualities of mind needed to give it highest significance and greatest value. Probably other planets have brought forth intelligent beings who equal or surpass man in their ability to understand and appreciate, but they are not evident in our solar system.

Although the Universe appears to have been set to evolve in a certain direction, its evolution has obviously not been directed by a cosmic mind with the foresight and power to guide it unerringly to the highest goal. Evolution has often blundered and gone astray, again and again destroying what it has accomplished and making fresh starts. But, with indomitable persistence, it has never ceased to strive, with the result that it has created much that is beautiful and admirable along with much that is hideous and repugnant. Because of its fumbling methods, its finest accomplishments have been bought at the price of immense suffering.

Man, one of the latest products of cosmic evolution, bears the marks of the groping process that shaped him. On the one hand, if we concede that a planet which lacks inhabitants who gratefully appreciate and try to understand it has not attained its highest value, we must further recognize that man begins to fill his planet's lack. On the other hand, it is clear that man still fills this lack most imperfectly and that, in consequence of the long evolutionary agony that molded him, he has many defects which make him dangerous not only to himself but to the planet which supports him. Although man has attributes which enable him to play an essential role on his planet, he was certainly not designed by a mastermind to fill this role to perfection.

Is it not paradoxical that the animal best equipped, in

senses and mind, to enjoy his planet's beauty should be, of all this planet's multitudinous species, the most destructive and the greatest threat to its continued prosperity? This paradox is not altogether inexplicable. To carry on his unique function, man needs a capacious mind, which in turn requires a fairly large brain, to bear and nourish which a fairly large body is necessary. The larger the animal, the heavier the demands it makes upon its environment; big animals tend to be destructive, to vegetation, to the soil, or to animals of other species on which they prey. Moreover, to develop his mind, increase his understanding, express his aesthetic impulses, and transmit his achievements from land to land and generation to generation, man needs many appurtenances: books, scientific apparatus, artists' supplies, means of communication and travel. He needs buildings to shelter himself, his libraries, his scientific equipment, and much else; he needs clothing and furniture for his comfort. Man leans more heavily upon Earth's productivity than any other animal, regardless of size.

In addition to all that appears indispensable if man is to carry on his peculiar function of understanding and appreciating his world and thereby increasing its value, he makes many unnecessary demands upon his planet's bounty and oppresses it in indefensible ways. His greed and craving for superfluous and even harmful luxuries enlarge immeasurably the burden he lays upon nature's largess. By his primitive hunting instincts, no longer necessary for his survival but difficult to outgrow, he afflicts his brother animals. His insensate wars are as destructive to the planet as to his own cities. By his inability to control his reproduction and keep his population adjusted to its environment, he threatens to overwhelm Earth with his excessive numbers.

This has created a perplexing situation. Of all the inhabitants of an exceptionally favored planet, we are best equipped to enjoy, understand, and appreciate it. However, the same evolutionary process that gave us superior endowments burdened us with appetites and passions that make us injure or destroy what we most need. To correct this perilous situation, several approaches are helpful. In the first place, we must intensify our awareness of the uniqueness of our planet. The space exploration of recent years has done much to make thoughtful people recognize how exceptional in the solar system their natal planet is—for this alone, such exploration is worth every cent of the billions of dollars that it has cost. Second, we must discover the role that most befits beings endowed as we are on a planet such as ours—a matter of insight and philosophical interpretation. Third, we must, by education and culture, deepen our understanding and intensify our appreciation of the good and the beautiful. Finally, we must develop the moral strength to control or subdue all those impulses and passions, foisted upon our stock by the long evolutionary struggle, that detract from our ability to fill our proper role and make us perilous to the planet that we should love and protect.

Our planet stands at the crossroads. As far as we can tell, it has never, in the billions of years since the first rudiments of life arose in the primal seas, been in such a critical situation. If man, its dominant animal, makes a widespread, sustained effort to develop his peculiarly human qualities of understanding, appreciation, and responsibility, it may move steadily toward fulfillment. If, on the contrary, he permits his highest attributes to be overruled by greed, hostility, and blind reproductive drives, the abundance and quality of Earth's vegetable and animal life, including man's own, will steadily decline, and it may never realize the promise that man's dawning spiritual characters brought to it.

When astronomers have told us the extent of the Universe and estimated the number of its galaxies and stars, when physicists have measured the mass of the hydrogen atom and the charge on the electron, when geologists have determined the age and structure of Earth, when biologists have mapped the DNA coil that controls living bodies and classified all the plants and animals—when we have all this information and much more, the most searching questions of all, of interest not only to philosophers but to every thoughtful person, may remain unanswered. What is the worth of this immense Universe and all that it holds? What does it contain to give it value? What would be the loss if it were instantly annihilated? To such penetrating questions objective science, within its self-imposed limits, can give no answers—for science is concerned with facts and measurements, not with the values that enhance existence and make life precious.

I gaze upon the stars and wonder what kinds of creatures might exist on the planets that probably orbit around many of them. I watch the birds flitting through the trees in my garden and wonder what thoughts or feelings occupy their little heads. In neither case can I do more than speculate vaguely; science does not enlighten me. Yet enlightenment in either sphere would immensely deepen our understanding of this mysterious Universe, of which positive science, for all its marvelous discoveries, gives us at best a one-sided view, since much that certainly exists is inaccessible to its procedures. This limitation is so frustrating that we continue to conjecture and to search, clutching hopefully at even the most tenuous clues. Book after book appears, soberly assessing the probabilities of life on planets within or beyond our solar system, while writers of science fiction give us fantastic pictures of extraterrestrial beings. On the other hand, we try increasingly to penetrate the minds of the diverse animals that share Earth with us, to understand their psychic processes and to communicate with them. Whether we seek conscious life on distant planets or try to assess its quality in the creatures around us, the results to date are lacking, inadequate, or unconvincing. Nevertheless, in both cases the effort is immensely valuable. It keeps our minds open to the exciting possibilities to which those of too much of mankind have too long been closed, and, with intelligent persistence, it may at long last yield the enlightenment that we eagerly seek.

Although the minds of the creatures that surround us often seem as inaccessibly remote as the possible inhabitants of the satellites of distant stars, when we search in this direction for evidence that beings capable of enjoying their existence and thereby adding value to the Universe are widely diffused through it, we do not grope so vaguely. We are certain that these animals exist, and only the most stubborn skeptic will deny that at least the more highly organized of them feel and are stirred by emotions, even if they are incapable of rational thought.

A creature may contribute to the total worth of the cosmos by enjoying its own life, or by contributing to the enjoyment of other beings, or in both these ways. We know that we are capable of enjoyment and that our lives are enhanced by the presence of other people with amiable qualities, so that we are certain that humans are, or are capable of being, both generators and enjoyers of values. Our delight in animals that are beautiful, graceful, songful, or skillful builders leaves no doubt that they generate values, and we strongly suspect that at least some of them, finding satisfaction in their own lives and activities, also enjoy values. When we turn to plants, which not only support our lives but embellish them in no small measure, we are more doubtful whether they derive any satisfaction from their growth in the sunshine, although it would be dogmatic to assert that they are incapable of feeling. A comprehensive survey of the ascent of life should give attention to both aspects of value—the ability of each organism to enjoy its own existence as well as its capacity to enhance the lives of surrounding creatures. Obviously, we are on firmer ground when we consider how the various forms of life enrich ours than when we try to assess their capabilities for enjoyment, yet to give our reasons for believing that they are, in some measure, sentient beings that find satisfaction in existence is certainly no more futile than to speculate about life on planets so distant that the most powerful telescopes fail to reveal them.

As a naturalist engaged for over half a century in the study of animal and vegetable life in tropical America, with fairly wide reading in philosophy and religion, I have pondered these questions long and earnestly. In this book I try to answer them, in the belief that to have answers, even answers tentative and subject to revision, should help us develop a view of the Universe, and our place in it, that would fortify our resolution to solve our problems and play our proper role on this exceptionally favored planet. Perhaps, after all, the Universe is what we and all its other parts, especially those of most advanced organization, can make of it.

The following chapters survey the evolution of life from a viewpoint different from that of the innumerable volumes which expound evolution's methods or trace the geologic history of plants and animals. Concerned less with methods and structures than with values, this book attempts to answer another question: what has evolution accomplished to make it worthy of our loyalty? Only to the measure that it raises the value of existence in its widest aspect, making it more satisfying and desirable, does evolution deserve our approval rather than our condemnation as a harsh and wasteful process. The single chapter that reviews widely accepted explanations of evolution compares it to a game of chance, by which life has had to gamble its way upward in the absence of intelligent, compassionate, foresighted guidance.

After calling attention to the uniqueness of Earth and its fitness to support life, we survey the stages of life's ascent from its simplest beginnings, giving special attention to the values, actual or potential, that slowly emerge. Then we pass to man, his exceptional endowments, physical and mental, his contributions to a planet where, for all his imperfections, he fills an ideal need and raises creation to a higher level. The final chapters suggest ways of viewing our relation to the Whole of which we are parts that should elevate our self-respect, strengthen our loyalty, and make us feel less alien. I hope that, by drawing attention to the importance of our planet and the splendid gifts that evolution has given us by so vast an effort, this book will help dispel the debilitating gloom that is settling upon humanity in a distressing age, so that with fresh courage and vigor we may support life's ascending course.

LIFE ASCENDING

1. The Fortunate Planet

As matter gravitates toward matter, straining always to come closer to what is most like itself, so spirit seeks spirit and mind thirsts to communicate with mind. Since he became thoughtful, man has never been reconciled to the possibility that this vast Universe is devoid of other beings with minds somewhat like his own, susceptible to pleasure and pain, aspiring and fearing, trying intelligently to improve their condition, striving to understand the world in which they live. He has attributed human thoughts and motives to the animals around him; he has peopled forest, field, water, and sky with a host of invisible spirits; and above all he has sought God, the Supreme Mind that knows all, that understands man and perhaps helps him in his difficulties.

Moreover, since astronomers demonstrated that Earth is only one of a family of planets circulating around the Sun, that the countless multitudes of stars are more distant suns which may well shine upon planets of their own, thoughtful men have been asking whether some of these planets support life, especially intelligent life. For some years, scientists have been trying to pick up signals that might emanate from a technologically advanced civilization on a satellite of one of the nearer stars. As I write, the Pioneer spaceship, speeding toward Jupiter, bears a plaque that, if all goes well, it will carry beyond the solar system, with the remote possibility that it will, after thousands of years, be picked up by inhabitants of another planet. From its inscription, they might conclude that intelligent beings in a far distant world have been trying to communicate with them.

The assumption underlying these recent, costly attempts to communicate with intelligent beings on other worlds is that, whatever strange bodily forms they may have evolved in environments different from ours, their minds work much as do our own, and they are no less eager than we are to understand the Universe, no less curious to know whether they are the only rational creatures in it. That all intelligent minds are somehow akin, that reason is the same wherever it arises, that all rational beings have somewhat similar aspirations is, I believe, a fair assumption.

Let us imagine a spaceship from a planet of some distant star carrying toward our solar system a party of scientists equipped with telescope, spectroscope, and an array of other sensitive instruments. These astronauts are eager to explore another planetary system and compare it with their own—above all, to learn whether life has arisen anywhere in it. Before undertaking a landing which, if imprudently made, may be irreversible, they guide their ship inward toward the Sun while they survey its attendant bodies.

First the astronauts approach Pluto, a planet about one-fourth the diameter of Earth and so far distant that it remained unknown to mankind until 1930. From a Sun so remote that it appears hardly larger than a star, Pluto receives little radiation to dispel the chill of empty space. Its temperature of about 380 degrees Fahrenheit below zero does not make it promising as an abode of life and forbids any attempt to land upon it.

Nine hundred million miles nearer the Sun, the explorers come to Neptune, a huge planet 28,000 miles in diameter, with two attendant satellites. But it is only a little warmer than Pluto; its dense atmosphere, rich in methane, would suffocate organisms that breathe oxygen; and it does not invite a visit. The next planet, Uranus, is a thousand million miles closer to the Sun than Neptune and, accordingly, somewhat warmer, but still at a temperature of about minus 350 degrees Fahrenheit at the cloud tops. Its imposing mass, 30,000 miles in diameter, is enveloped in an atmosphere of hydrogen and methane, and five satellites circulate around it.

As they approach the next planet toward the Sun, Saturn, our explorers are fascinated by its extraordinary rings, unlike anything they have seen elsewhere in the cosmos. Many thousands of miles wide from inner to outer edge but only a few miles thick, the flat rings are composed of millions of tiny, icy moonlets or satellites, orbiting a planet 75,100 miles in diameter. One of the planet's fifteen larger satellites is nearly as big as Mars. No prudent astronaut would attempt to reach Saturn's surface, hidden by a dense, frigid atmosphere of hydrogen and helium. It appears to be another lifeless world.

While still a long way from Jupiter, the astronauts become aware that its immense mass, 318 times that of Earth, is deflecting their craft strongly toward its surface. To avoid being drawn down to a planet with a gravitational field so intense that they could never escape, they increase their velocity to shoot by at a safe distance, while they contemplate the imposing spectacle of a body 88,700 miles in diameter, surrounded by thirteen moons of varying sizes. Two are larger than the planet Mercury, but what most stirs the wonder of the party's astronomer is that the four or five outermost circulate rapidly from east to west, in a direction opposite from that of all the others and of the rotation of the swiftly spinning planet itself. Can they be asteroids that had floated into Jupiter's powerful gravitational field and been captured by it?

As the astronauts penetrate more deeply into the solar system, they find each succeeding planet somewhat warmer than the last. But Jupiter, 485 million miles from the Sun, is still forbiddingly cold, with the temperature in its cloudy upper atmosphere around minus 220 degrees Fahrenheit— about that of liquid air. Composed largely of hydrogen and helium, with some ammonia, methane, and water vapor, this atmosphere is evidently so many thousands of miles thick that it prevents any view of the planet's solid surface, if there is one. Jupiter's enormous gravitational pull is caused by its huge size rather than by its density, only one-third greater than that of water. Whatever rocky core it might have must be relatively small; but toward the center of such an enormous thickness of gases each would assume the liquid, then the solid, state, unless all became much hotter. Possibly a tremendously deep ocean of compressed gases envelops a central mass composed of iron and other heavy elements. The varying shades of the planet's atmosphere suggest that it is subject to great turbulence, with terrifying storms of lightning and thunder. The incautious astronaut who ventured too near the gigantic planet would be trapped forever in a cloud of frigid, lethal gases. No organism that depends upon water, oxygen, and moderate temperatures could flourish there.

As our astronauts speed away from the imposing but forbidding planet, they speculate upon the possibility that some alien form of life might exist upon it. The organisms on their home planet, as on Earth, are composed largely of carbon, hydrogen, oxygen, and nitrogen, and their vital processes require an aqueous medium. Nevertheless, the essence of life is not so much a certain chemical composition as it is the capacity to grow, to reproduce, to react in subtle ways to a great variety of stimuli, and—certainly in its higher forms and most probably throughout its whole range—to feel more or less acutely. On their own planet, the travelers are familiar with certain very humble organisms whose chemistry is quite different from that of the dominant plants and animals.

If life arose in ammonia rather than in water, it could flourish at much lower temperatures, because ammonia has a much lower freezing point. Moreover, ammonia is a versatile chemical, capable of entering into many complex compounds, including some closely analogous to the proteins and nucleic acids indispensable to familiar creatures. Carbon for those compounds would be available from the methane so plentiful on Jupiter, if not from the traces of carbon dioxide present there. The tremendous pressure even at the surface of the hypothetical Jovian sea would be no more inimical to life than the smaller, but still enormous, pressure in the depths of Earth's oceans. Perhaps, too, ammonia-based organisms could float in Jupiter's dense atmosphere, growing and reproducing without ever resting on a solid or liquid surface. It would certainly be dogmatic to deny that beings which might be called alive could exist on Jupiter or even on planets still farther from the Sun, but they would be creatures unimaginably different from any that we know.

While discussing this intriguing question of alternative chemistries of life, our astronauts enter the broad belt of asteroids between Jupiter and Mars. Before long, they realize that they are in a zone perilous to spaceships, for it is traversed by innumerable bodies of the most various sizes, each a miniature planet circulating around the Sun in an independent orbit, with a velocity intermediate between that of Mars and that of Jupiter. If the astronauts joined the procession at the prevailing speed, they would be in little danger, but while traversing the zone from side to side they are exposed to devastating impacts. They hold their breaths in terror as a mass of naked rock, the size of a house, approaches them at a terrific speed; but, by jogging their vehicle off its course with a short blast of a lateral jet, they escape by a few yards. They pass close enough to the greatest of the asteroids, Ceres, to scrutinize it with their scopes; and they become convinced that this mass of rock and metal, nearly five hundred miles in diameter, has a gravitational field far too weak to hold an atmosphere and that its rough surface, deeply pitted by the impact of meteorites, supports no visible life.

As the globe of Mars expands before them, the astronauts feel a mounting glow of anticipation. Here, at last, is a planet with a visible surface! Before long they notice that the sphere, of a generally reddish hue, bears opposite caps of white, evidently polar ice, such as they knew at home.

5

But when their instruments disclose that the exceedingly thin atmosphere is composed mostly of carbon dioxide, with only traces of water vapor and oxygen, they suspect that the white caps are largely frozen carbon dioxide, known on Earth as dry ice.

The travelers scan Mars' surface intently, but they detect no glint of standing water. Yet its absence is puzzling, for as they approach nearer they see that the planet's extremely broken surface, with mountains far higher than any on Earth, is marked by the craters of volcanoes that must have emitted much water vapor, along with other gases. Moreover, they detect gullies and deep canyons which could hardly have been formed otherwise than by the prolonged action of voluminous rivers. But they find no trace of the long, straight canals which the American astronomer Percival Lowell, wrongly interpreting sketches made by the Italian Giovanni Schiaparelli, believed that he discerned on Mars. Lowell inferred that these channels were dug by intelligent beings, to conduct water from melting ice caps to cultivated fields at lower latitudes, thereby delaying the extinction of their civilization as their planet slowly dried.

Mars is everywhere reddish, yellowish, or gray, with no broad expanses of green to indicate flourishing forests or meadows. Although its scanty water and oxygen might support a little life, this is evidently of a quite humble order, perhaps corresponding to the lichens and mosses that grow on exposed rock and barren ground on more hospitable planets. A darker zone around the white cap on the pole which happens to be turned toward the Sun suggests that water from melting ice supports a more luxuriant growth. Whatever grew on Mars would need to be extremely hardy, able to withstand sunlight with its ultraviolet radiation only slightly filtered out by the thin atmosphere, as well as temperatures that, near the equator, may fall from about 50 degrees Fahrenheit while the Sun is directly overhead to minus 130 degrees during the night—a range about as great as that from melting ice to boiling water.

Just as the topography of Mars points to the former presence of more abundant water, so the redness of its deserts suggests that oxygen was once more plentiful, for this color is due to the oxidation of minerals. Why, then, are water and oxygen now so scarce? A clue to the puzzle is provided by the small size of the planet, little more than half the diameter of Earth, which, coupled with its lower density, gives it a mass only slightly over one-tenth of Earth's, while the force of gravity at its surface is about two-fifths that at the surface of Earth. Its escape velocity—the speed that an atom or a rocket must attain to break free of its gravitational field—is only 3.1 miles per second, while that of Earth is 7 miles and that of Jupiter 37 miles per second. At this low escape velocity, atoms and molecules, especially the lighter ones, are constantly flying off into space. Mars is currently losing its water supply at the estimated rate of 100,000 gallons per day. Evidently in past ages the planet was more generously watered, and with an atmosphere richer in oxygen it could have supported more abundant life, traces of which may be preserved as fossils beneath its deserts. Mars, as Sir H. Spencer Jones wrote long before the days of space probes, is a planet of spent life. Perhaps it presents a picture of what Earth will be like many millions of years hence, after most of its richer atmosphere has been dissipated into space.

While the astronauts continue to orbit Mars, intently studying its bleak landscapes, they become less and less visible. Violent winds have sprung up in its thin atmosphere, raising from its naked soil huge clouds of yellowish dust that obscure vast areas of its terrain. Leaving desolate Mars, the travelers are glad that they had not descended upon the barren planet, for such gales might wreck their spaceship.

The explorers next set their course toward Earth, then nearing conjunction with its neighboring planet, and visible in their telescope as a thin crescent of light, like a new Moon. Slowly the crescent grows larger, as they race toward it at a velocity not yet attained by man's spaceships. To avoid impact, they veer aside, with the intention of orbiting around Earth at a reduced speed while they survey it. When they come near enough to distinguish details, Earth's dark side is turned toward them; but the clouds along the disk's eastern margin glow from pole to pole with the colors of dawn, crimson and roseate and orange, a glorious display that seems to welcome these visitors from afar.

As their slowing spacecraft swings into orbit and more and more of Earth's sunward face rises into view, the astronauts press close to their viewing ports, enthralled by the prospect unfolding before them. In long, irregular, curving bands, like icing poured over a huge cake by sweeping movements of a gigantic hand, white clouds stretch for immense distances over the planet's face, with here and there smaller fleecy masses between the bands. The clouds cover nearly half the visible Earth. Between them blue, island-studded seas gleam in sunshine. The continents are variegated with large expanses of green that proclaims flourishing vegetation, yellow or brown deserts, mighty mountain ranges with snowcapped peaks, and great rivers winding like bands of polished steel through the plains. The poles of this colorful sphere are white with huge accumulations of snow and ice. The northern hemisphere is greener, for it is now the northern summer, but the southern shows by far the greater expanses of water. Although still too high to detect animals and uncertain how to interpret certain strange features that could have been made by intelligent inhabitants, the astronauts have little doubt that this favored planet abounds with life. In some ways it resembles that from which they came, but it is even more beautiful, for it is better watered and illuminated by a brighter star. A touch of nostalgia sweeps over the voyagers as their thoughts rush back to homes light-years away.

Again and again they orbit Earth, trying to distinguish every detail by coming as close as they dare without attempting to land. Certainly this planet must be explored from the surface, but before landing they wish to survey the two remaining planets of the solar system. Perhaps the next one will be even more inviting!

Half in sunlight and half in shadow, Venus shines beautifully as the astronauts approach, after describing a wide arc through space. With high anticipations, they reduce speed and swing into orbit around the planet; but, as they peer eagerly downward, they can see nothing of its surface. Here is no transparent atmosphere, like Earth's, but an extremely dense, murky envelope, heavy with dust, through which their vision cannot penetrate. When their instruments reveal that the Venusian atmosphere is composed chiefly of carbon dioxide and sulfuric acid, they conclude that, like the glass of a greenhouse, it would trap the rays of a Sun that shines fiercely upon it and create very high temperatures near the surface.

They could not know that, a short while before their visit, a Russian space probe revealed a surface temperature of about 900 degrees Fahrenheit and an atmospheric pressure nearly a hundred times as great as that at sea level on Earth, but they wisely decide not to land. Before they could reach the surface, their spaceship would be crushed and they themselves burned to cinders. Certainly no life based upon unstable protoplasm could exist upon the surface of this hot, Earth-size planet. Possibly light organisms might live and reproduce in the cooler upper layers of the heavy atmosphere, floating continuously, in perpetual danger of being cooked if they sank too low; but this is a conjecture that the astronauts are not prepared to investigate.

Mercury, hardly half the diameter of Venus, appears too small to retain an atmosphere. Its side turned toward a Sun only 36 million miles away becomes intensely hot; while the dark side, with no blanketing layer of gases, promptly radiates its heat into empty space and falls many degrees below zero. So inhospitable to life does its barren, densely crater-pitted surface appear that the astronauts are content to view it from afar. After a final swing around Venus, they increase their speed and shoot off toward Earth, which shines in the distance like an exceptionally brilliant star.

Their long exploration of the solar system has convinced these strangers that Earth is by far the most favored of its nine planets. Situated neither too near nor too far from the Sun; large enough to retain an abundance of air but not so massive that its atmosphere is excessively dense; with a liberal supply of oxygen mixed with more sluggish gases to moderate its too intense chemical activity; with much water; with immense expanses of fertile land rising above heaving oceans; with a mantle of lovely vegetation protecting its soils, constantly renewing its supply of life-supporting oxygen, and providing abundant nourishment for animals— with these manifold advantages, the fortunate planet has variety, beauty, and interest equaled by no other celestial body known to the voyagers.

With mixed emotions, the astronauts prepare to land upon this inviting but still untested world. Their knowledge of the Universe is broad enough to assure them that Earth is composed of the same elements as their own planet, perhaps mingled in quite different proportions, but beyond this they hardly know what to expect. Doubtless they will meet a vast array of utterly strange life forms, as fascinating as they are bewildering. They wish to compare Earth's geologic history with that of their home planet. They hope to meet intelligent beings who can, probably with much difficulty, communicate with them, perhaps throwing fresh light on problems that have long perplexed the scientists and philosophers of their own civilization.

Opposed to these enticing prospects are insidious fears. Would this untried environment contain, in air, soil, or water, substances lethal to strangers? Would the atmosphere at the planet's surface be too rare, or too heavy, for them to breathe? Would they encounter pathogens to which they had developed no resistance? They wish to bring a message of peace and friendship and assurance that intelligence is widely, if thinly, diffused through the Universe; but would the dominant inhabitants of this unknown planet rudely reject the advances of beings so strange in aspect, perhaps injure or kill them as dangerous invaders?

It never occurs to the astronauts that, if they touch down in a densely populated district, they may be crushed or suffocated by multitudes of curious bipeds crowding around them, incredulous of their distant origin.

2. The Importance of Earth

The dominant inhabitant of the fortunate planet is its most social animal. This is not because man crowds into huge cities with millions of inhabitants. Some of the social insects, especially among termites and ants, live in comparably populous communities, packed together more tightly than most humans would find tolerable. Indeed, many urban dwellers exist in solitary loneliness amid the throng. We are the most social animals because we crave a companionship wider than that which our own species affords. We cultivate the society of our brother animals, of plants, of hills and brooks, of the stars above us and the ground beneath. For many of us, this wider society is as necessary for spiritual health as is that of our fellow men. Sometimes a person is called unsocial because he spends little time with other people; but, if you investigate his social circle, you may find that it includes a substantial segment of the natural world or, perhaps, minds that lived long ago. He is no less social than those who demand the constant presence of human companions; his sociality takes a different direction. The more advanced his spiritual development, the more social he becomes, the wider and more diverse the fellowship he seeks. He needs a society much wider than that of mankind.

It is no accident that we are social; the very atoms of which our bodies are made are social beings. They display both an undiscriminating and a more selective sociality. The former manifests itself as gravitation, which causes them to aggregate in bodies of which the tightness depends solely upon their mass rather than upon their kinds. The discriminating sociality is sometimes called chemical affinity. For their closest unions, atoms will have partners only of certain kinds, with which they join in patterned structures known as molecules. In favorable conditions, as in a drying solution or a cooling magma, the molecules line up, rank upon orderly rank, to form crystals, which may endure for long ages and are often so beautiful that they are called gems. In other situations, as in the tepid seas, rich in solutes, on a cooling planet, the molecules may combine in aggregates, far more complex but much less stable, that have the peculiar ability to reproduce themselves, thereby becoming the forerunners of life.

This social urge that causes atoms to arrange themselves in patterns of ever increasing amplitude, coherence, and complexity is what sets the world process in motion. It impels matter ever forward, gathering it into suns and planets, then covering the more favored of these planets with living beings, some of which finally become able to feel strongly, to think, and to respond to the beauty that surrounds them. We might call this process harmonization, because it orders some of the originally diffuse materials of the Universe into harmoniously integrated patterns. One particular phase of this grand movement is organic evolution, in which harmonization is complicated, and often masked, by the complex and frequently destructive interactions between its living productions.

To realize their creative potentialities, the social atoms require rather special conditions. In the first place, they must be neither too scattered nor too condensed. Astronomers estimate that about half the matter in the Universe now floats in vast clouds of gas and dust in the interstellar spaces. In such a rarefied medium, development cannot proceed far; simple molecules, and perhaps small crystals, may arise, but not molecules of a complexity remotely approaching those of living organisms. By shrinking to a very small fraction of their original size, heating up as they condense, the gas clouds may, however, form stars. As these stars grow old, the matter in the interior of the heavier of them becomes very dense. Under the tremendous pressure of the overlying mass, the very atoms are crushed, stripped of their electrons, so that their bare nuclei become closely packed together. It has been estimated that in the interior of white dwarf stars matter is so compacted that a handful of it, if it could be brought unaltered to Earth's surface, would weigh hundreds of tons. Even at the center of the larger planets, such as Jupiter, matter may lie crushed and degraded beneath the superincumbent mass, unable to display its constructive powers.

Second, to realize its creative potentialities, matter must be neither too hot nor too cold. Neither complex molecules nor crystals can exist at very high temperatures. The Sun is a sphere of incandescent gases 865,000 miles in diameter, with even its metals volatilized. At the other extreme, as the temperature approaches absolute zero (minus 459.6 degrees Fahrenheit), the movements of atoms become so sluggish that chemical reactions are retarded and finally cease altogether. Very low temperatures are favorable for preserving complex structures but not for building them. At high temperatures, on the contrary, the extremely mobile atoms may momentarily fall into the most diverse configurations, but they cannot preserve them. Only at moderate temperatures can complex molecules arise and persist. The narrow range between the freezing and boiling points of water appears to be the most favorable for the growth and preservation of complex molecular patterns.

It follows from this that to exhibit its constructive power, especially to assume the living state, matter must lie at or near the surface of a planet neither too near nor too far from the star that warms it. Mercury and Venus are too close to the Sun, Jupiter and all the outer planets too remote. Earth and Mars revolve about the Sun in the optimum zone, with our planet more favorably situated. It does not occupy this privileged position because it is ours; on the contrary, we are here because of its fortunate location.

Moreover, for life to arise and evolve to high levels, a planet must be neither too small nor too massive. If too light, it will lose most or all of its atmosphere and have little or no surface water, as has happened to the Moon and is happening to Mars. If too massive, the planet will retain too dense an atmosphere which will screen the Sun's energizing rays from its surface. In this respect, too, Earth has been particularly favored. Intermediate in size between the smallest and the largest members of the planetary family, intermediate in distance from the Sun between the nearest and the most remote, it has become the fairest, most fertile of them all.

Explorations of the solar system, first with telescope and spectroscope, more recently with space probes, permit us to assert with growing confidence that this is at present the only planet covered with a living mantle of vegetation, the only one on which flowers bloom, butterflies float in sunshine, birds sing, silvery fishes glide through limpid water, and intelligent, appreciative inhabitants paint or photograph its landscapes, celebrate its beauty in verse, and try to fathom its origin and destiny. With the exception of Mars, scorching Mercury, and possibly freezing Pluto, it is probably the only planet from which surface-dwelling creatures could view the starry heavens and wonder at the immensity of space; all the others appear to have a densely obscuring atmosphere.

The importance of Earth is now becoming apparent. This is due not to its size, for even in the solar system it is small, occupying but an infinitesimal fraction of universal space, or to its composition, for it is made up of elements which spectroscopy reveals to be widely diffused through the cosmos. On the contrary, its importance may be ascribed to the very fact that it is made of materials common to all the stars and planets. Everywhere matter has the same potentialities. Elements present in Neptune, the Sun, Arcturus, or the farthest galaxy could make a snow crystal, a tree, a bird, or a man if put together in the proper pattern in a suitable environment. But only in certain relatively small, favorably situated bodies can the stuff of the Universe carry to a high level of development the creative impulse that seems to pervade it everywhere; only at certain special points in the vastitude of space can matter realize its latent potentialities. In our solar system, the chief of these points is the planet Earth.

Earth, then, may be regarded as an expression point of the Universe, a privileged spot where the creative energy reveals what it can accomplish. Its value must be reckoned not by its size but by its key position in a planetary system spread through some 8,000 million miles of space. Yet we have no reason to conclude that even here harmonization has attained its highest, most perfect expression. We hope that it can fashion rational beings far superior in many ways to our present imperfect selves, that it can populate a planet with animals that dwell in harmony instead of fighting and devouring each other, as on Earth. Yet all our present evidence points to the conclusion that here the formative power of the Universe has achieved an expression far more adequate than any elsewhere in our solar system. Earth is a unique stage where the creative forces present an exceptionally fine performance.

To ancient man, Earth was the center of the Universe, the reason for its existence. The stars and planets were but sparkling adornments of the night sky or, in the view of certain Classical philosophers, blessed gods dwelling serenely far above our mundane turmoil. But, as early as the fifth century B.C., Philolaus, of the Pythagorean school, advanced the view that this terrestrial sphere, along with an invisible counterearth and all the visible celestial luminaries, revolved around a great fire that occupied the post of honor at the center of the cosmos. About the same time, Democritus and his school taught that Earth is one of many worlds, all formed by the fortuitous concourse of the restless atoms. As early as the third century B.C., the Greek astronomer Aristarchus of Samos proposed a cosmic system in which Earth and the other planets then known circulated around the central Sun.

Since the mechanics of such a system had not yet been elucidated, astronomers were not prepared for its acceptance. For many centuries, they continued to use the system of Ptolemy, with its complicated scheme of cycles and epicycles to explain the movements of the planets, now advancing among the fixed stars, now lagging behind them, in a manner that annoyed certain Greek philosophers, who thought that they should march forward at an even pace, like gentlemen. Not until early modern times did the observations of Copernicus and Tycho Brahe, the calculations of Kepler, and the mechanics of Newton place the heliocentric theory of the solar system on such firm ground that it has won almost universal acceptance.

Ecclesiastical opposition to the heliocentric theory was strong, for it contradicted the Biblical account of creation, dethroned Earth from its central station in the Universe, and made it more difficult to believe that the world was made for man, the chief preoccupation of his Creator. As ever better and more powerful telescopes were pointed toward the sky, Earth continued to shrink in relation to the Universe. Uranus, Neptune, and, after a long interval, Pluto were added to the six planets that had long been known, quadrupling the span of the solar system. Our Sun was found to be only one among a hundred thousand million stars in the Milky Way and by no means the largest and most brilliant. Our galaxy itself turned out to be but one among many millions of similar star groups scattered widely through a Universe so vast that light, traveling at 186,000 miles per second, requires thousands of millions of years to traverse it. Each successive amplification of the known Universe by astronomical discoveries seemed further to humble our tiny planet, reducing it to almost absurd insignificance. In this judgment, thoughtful men everywhere have tended to concur with the churchmen who stubbornly resisted the innovations of Copernicus, forced Galileo to retract his heretical pronouncement that the Earth moves, and burned Giordano Bruno at the stake for proclaiming the plurality of worlds.

This widespread view of the shrinkage in stature of Earth, and with it of the planet's dominant inhabitant, man, in consequence of astronomical discoveries, springs from misplaced emphasis. Rightly viewed, Earth is more important now than it could be in the little, circumscribed Universe of our ancestors, since it is the expression point, the most highly evolved planet, of a solar system far vaster than they suspected. It reveals, as no other of the nine planets appears to do, what the materials of which all are composed can achieve in beauty, in awareness, and in significant existence when they find a proper environment—one neither too hot nor too cold, neither too rarefied nor too dense, and sufficiently stable through long ages. Our planet is a revelation of Being's potentialities for joy and exaltation, as well as for sorrow and terror.

Although unique in our solar system, doubtless Earth has counterparts in the Universe at large. But, if stars other than our Sun have planets, they are not big enough and luminous enough to reveal their presence in our telescopes; we can only speculate about their existence and the possibility that they support life. Estimates of the number of planetary systems beyond our own are strongly influenced by theories of the origin of such systems.

These cosmogonic views are of two general types, which for brevity we may designate as interference hypotheses and contraction hypotheses. According to the former, which appear to have originated in the fertile imagination of the prolific French naturalist the Comte de Buffon, the planets were drawn out of the Sun by a collision or a near collision with some other star. Approaching at a tremendous speed, the star may have plowed through the surface of the Sun, splashing out huge quantities of gaseous matter, which reached tremendous distances before the Sun's gravitational attraction stopped the flight of some of it and drew it into orbit. More probably, the intruding star failed to strike the Sun but shot by, perhaps at a distance of a million miles, which would be close enough to draw up a great protuberance or tidal wave, similar to that produced in Earth's oceans by the attraction of the Moon and Sun, but immeasurably more voluminous. One crest of the wave would reach up toward the passing star, while a lower bulge would arise on the Sun's opposite side. If the passing star were more massive than the Sun, its attraction might tear away much of the crest directed toward it, scattering the material through space as it sped away. Some of the matter detached from the Sun, whether by impact or by gravitational attraction, could in the course of time condense to make the planets, their satellites, and the asteroids.

Of the contraction hypotheses, the best known is that associated with the illustrious names of the German philosopher Immanuel Kant and the French mathematical astronomer Pierre Simon de Laplace, although it had earlier been adumbrated by the Swedish scientist and mystic Emanuel Swedenborg and the Englishman Thomas Wright. According to the nebular hypothesis, our solar system was originally a vast cloud of gases and dust which must have extended beyond the present orbit of the outermost known planet, Pluto. As gravitation drew the diffuse matter together, the cloud spun faster and faster, becoming flattened into a disk under the influence of centrifugal force. The rotation became so rapid that at the disk's rim this force balanced the gravitational attraction of the mass. When this happened, the peripheral material remained behind as a ring, while that within it continued to contract until a second ring separated from the cloud in the same fashion. Of the ten rings successively cut off in this way, nine somehow condensed to form the nine planets, while the tenth remained in a more diffuse state as the belt of asteroids. The residual central mass became the Sun. The planets' satellites were created by repetitions, on a much smaller scale, of the same process that gave rise to the planets themselves.

The origin of the solar system has proved to be one of the most perplexing of cosmological problems. Even if it were possible for planets to be formed in the manner proposed by the interference hypotheses, few could arise in this way, because their birth depends upon an accident that rarely occurs. The stars are separated by such immense distances that the probability of one passing close enough to another to detach material from it is exceedingly slight. If this is, in fact, how our solar system originated, only a very small minority of the stars would be likely to have such a system.

According to the condensation hypotheses, however, planets arise in the normal course of the evolution of a star, or at least of stars of a certain type, so that planetary systems may be quite numerous, not only in our Milky Way but in other galaxies. Not long ago, the nebular hypothesis was in disfavor among astronomers because it failed to account for the dynamics of the solar system. More recently, modifications and subsidiary explanations proposed by C. F. von Weizsäcker, H. Alfvén, G. P. Kuiper, Fred Hoyle, and others have made a condensation hypothesis acceptable once more, so that we can now believe that stars with planets are by no means a rarity. Yet given any view, the estimation of their number, and of the number that may support life, is fraught with tremendous uncertainties.[1]

The galaxy in which we live, the Milky Way, is estimated to contain about a hundred thousand million stars. A large proportion of these stars, possibly about half, are multiple, consisting of two or three or more huge, incandescent masses revolving around each other, held together by gravitation. A doublet or a triad probably results from a mode of star formation alternative to that which gives rise to planets. In any case, a planet associated with such multiple stars would trace a most complicated course, now far from the incandescent bodies and now near them, resulting in great fluctuations in temperature that would be inimical to life.

[1] For a review of this subject, see Jagjit Singh, *Great Ideas and Theories of Modern Cosmology* (London: Constable, 1961).

Of the single stars, which alone could illuminate a planetary system comparable to our own, the Milky Way might contain as many as 50 billion. Some are too hot and some too cool, some too young for life to have risen beneath their rays, others so old that life may have flourished and become extinct on their planets. As to the number of planets that different stars have, we can only guess. Some could have only one, others might have many more than our Sun. If the number of planets is great, some are almost certainly too near the star and too hot to support life, while others are too remote and too cold, as in our own system. Some are too small to retain an atmosphere, others so massive that their atmospheres are oppressively dense. Yet it is fair to assume, with Sir Harold Spencer Jones, that wherever conditions are favorable for life it will arise if enough time is available.[2]

Because of all these uncertainties, estimates of the number of planets on which life has arisen are hazardous and vary widely. P. H. A. Sneath suggested that "there are probably not less than 200,000 and not more than 800 million life-bearing planets in our galaxy."[3] Captain David C. Holmes placed the number at "more than 100,000."[4] But the cosmos contains many millions of galaxies, so that, even if we estimate the number of life-bearing planets in our home galaxy as improbably low as one thousand, the number of such planets in the Universe as a whole could be immense. And it would be surprising if some of these millions of inhabited worlds did not support beings as well endowed intellectually and spiritually as ourselves or even far surpassing us in many ways, although in outward form utterly different from ourselves.

It is highly probable, then, that we are not the only be-

[2]Sir Harold Spencer Jones, *Life on Other Worlds* (London: Macmillan, 1940; New American Library reprint).

[3]P. H. A. Sneath, *Planets and Life* (London: Thames and Hudson, 1970).

[4]Captain David C. Holmes, U.S.N., *The Search for Life on Other Worlds* (New York: Bantam Books, 1966).

ings who appreciate and try to understand the cosmos, who are grateful for the privilege of living in it. Evidently the Universe has many expression points, some of which may reveal what it can accomplish—in beauty, in joy, in spiritual fulfillment—more adequately than our own. The certainty of this would hearten us and encourage our efforts to improve ourselves and our world. Nevertheless, given any estimate, these expression points are very thinly scattered through space, and their presence, at immense distances from us, hardly decreases the importance of our own. Within a radius that must be measured in light-years, on no other planet has creation reached so high a level as on that of which we are the fortunate inhabitants.

Realization of the importance of our planet should make us value it as men too seldom do. In the growing concern for the terrestrial environment today, we detect a wide range of motives and attitudes. Some would treat Earth as a farm that should be made to yield as much as it can for human use, without in the long run diminishing its productivity. Its worth is measured by the number of people it can continue to support in a satisfactory degree of comfort. For others, Earth is too rare and beautiful, too richly diversified, to be treated merely as a means for satisfying man's prosaic needs. They regard it not as a farm but as a garden that is cherished, not because it yields food or wealth but because it uplifts the spirit with its loveliness. To be sure, unless Earth is made to produce enough food and other necessities to keep us alive and healthy, we shall not survive to appreciate its uniqueness. To awake to our privilege and responsibility to make this exceptional planet an ever more adequate expression of everything excellent that the creative process can bring forth would give human life fresh zest, new dignity, and immeasurably heightened significance.

3. The Sphere of Life

The materials that compose our planet surround a central core in concentric hollow spheres of decreasing density. Although the deepest mine shafts and borings hardly penetrate Earth's skin and fail by many miles to reach the central nucleus, indirect evidence, based largely upon the propagation of earthquake waves, indicates that it has a radius of about 2,150 miles and is composed largely of molten iron, with an admixture of nickel and smaller amounts of other metals. Near the surface, the iron has a temperature of about 9,300 degrees Fahrenheit.

Surrounding this central core is a mantle of basaltic rock about 1,800 miles thick, which, despite the high temperature of its deeper layers, behaves much like an elastic solid because of the great pressure upon it. The mantle is, in turn, enclosed by a crust of solid rock, chiefly granite under the continents and basalt under the oceans. The thickness of the granite varies from about twenty to forty miles with the height of the land, being greatest under high plateaus and mountain ranges, as is necessary to support their weight as it floats on the heavier underlying basalt like an immense iceberg in the water. Under the oceans, the basaltic crust is much thinner, perhaps only two or three miles in places.

The planet's crust is divided into about a dozen huge, continent-size plates, which with extreme slowness slide over the mantle, changing their relative positions. In the Pennsylvanian or Upper Carboniferous period, about 300 million years ago, Africa, South America, India, Australia, and Antarctica were in contact, forming one enormous land mass known as Gondwanaland. The New World continues to inch away from the Old, widening the Atlantic. Although the immense internal forces causing this movement are poorly understood, the line of separation has been identified as the rift along the middle of the immensely long, submerged mountain range that extends almost the whole length of the Atlantic Ocean. This undersea canyon in the Mid-Atlantic Ridge, the site of great volcanic activity, has recently been closely examined by American and French scientists in submersibles designed to withstand the tremendous pressure of the ocean's depths.

If Earth's surface were quite level, it would be uniformly covered by an ocean two miles deep, and we might recognize a hydrosphere enclosing its crust. However, the huge corrugations of the planet's surface have broken the continuity of the oceans, leaving large areas of land exposed, so that the water hardly forms a sphere. Next above the crust, then, comes the atmosphere, which embraces the planet in unbroken continuity. With increasing altitude, the density and temperature of this blanket of air change, permitting us to divide it into concentric zones, counterparts of those inside Earth. The lowest and densest region of the atmosphere, the troposphere, is about five miles thick at the poles and twice as deep at the equator. This is the unstable birthplace of all weather, with its alternating calms and storms, its waves of heat and cold. Despite fluctuating temperatures at any point, a general decrease of about 3 degrees Fahrenheit for each thousand feet of altitude continues until we reach the tropopause, the layer of thin, bitterly cold air that separates the troposphere from the stratosphere.

With a thickness of forty or fifty miles, the stratosphere is a region of gentle, steady winds, occasional light, mother-of-pearl clouds, and rising temperatures. As one ascends through this layer, the Sun beats down with an ever fiercer glare, while in other directions the stars shine out in a darkening sky. Still higher, where the air has become exceedingly thin, its atoms are electrified by ultraviolet rays and the impact of particles shot out from the Sun. This is the ionosphere, which lies about sixty to eighty miles above the ground and, reflecting the waves used in radio transmission as a mirror reflects the much shorter waves of light, bends them around the planet's curvature and enables us to hear programs from the opposite hemisphere. Here, surprisingly, the temperature rises to 212 degrees Fahrenheit, the boiling point of water, and occasionally much higher.

Above the ionosphere, the air becomes attenuated so gradually that it is impossible to say exactly where the atmosphere ends and outer space begins. Even at heights of hundreds of miles, the weak flickering of the aurora borealis and the incandescence of meteors indicate the presence of gaseous molecules, as thinly scattered as any in the most perfect vacuum that can be made in a laboratory.

The Extent of the Biosphere

In addition to the concentric zones based upon the planet's lifeless constituents, the presence of life permits us to recognize another, the biosphere. Although thinner than any of the foregoing zones, it is the most diversified and interesting, the meeting place of soil, water, air, and sunlight, the crowded stage where trillions of individuals—more than a million species of living things—interact and work out their destinies. The biosphere has two grand divisions, the marine and the terrestrial; its global continuity is broken only by the high, ice-covered, intensely cold interiors of the Antarctic continent and Greenland and the scattered mountains at lower latitudes that rise above the limits of vegetation. Even here, stray living things may occasionally be found.

To assign a precise vertical limit to the sphere of life is hardly feasible, but to do so even roughly will enable us to make some interesting comparisons. On land, life is most concentrated in vegetation, from the topmost layer of the

soil up to the tips of the plants. Although certain exceptional California redwoods rise to heights of 368 feet and
some Australian eucalyptus trees are nearly as tall, few
forests anywhere spread their canopies more than 200 feet
above the ground. Even in rich tropical woodlands, only
unusual trees rise so high. In arid lands, the scattered plants
are very much lower; and, over vast areas of far-northern
tundra, the stunted vegetation hugs the ground. When we
take into account, too, the great expanses of more or less
treeless grasslands and barren deserts, if we place the average height of vegetation over the continents and islands,
even before man destroyed such immense areas of forest, at
100 feet, this will be a liberal estimate.

Insects fly above the grasstops of meadows and even over
the roofs of high forests. Sometimes they are carried to
great heights by updrafts, and this frequently happens to
newly hatched spiderlings dispersing on their gossamer
threads. Most birds of the tropical forests remain down
amid the trees; but swifts, swallows, and kites frequently
circle above them, catching volitant insects, as do vultures
and hawks scanning the ground or vegetation for dead or
living food. Migrating birds travel at considerable heights,
probably mostly between two and six thousand feet above
sea level when traversing low-lying land or water, very
much higher when crossing great mountain ranges. But all
these high-flying creatures reproduce down amid the vegetation, where all their food originates. This is their true
home; they are only transients at higher levels. This situation perplexes us when we attempt to determine the upper
limit of the biosphere. We might take it to coincide with
the upper limit of vegetation, but let us be generous and
add 150 feet. If we set the upper limit of the terrestrial part
of the biosphere higher, we shall have to do the same for
the marine portion. Insects and spiders are carried by the
winds high over the oceans—Darwin told how spiderlings
alighted on the rigging of the *Beagle* far out at sea.

Downward, the biosphere extends into the ground at least

as far as roots penetrate. In poorly drained soils and those with a high water table, root systems tend to be shallow. In arid lands with well-aerated soil, plants send their roots deeper, searching for water; even herbs may penetrate the ground for three or four yards. The soil is inhabited by moles, gophers, and other mammals; many insects, especially while in their larval stage; earthworms; and huge populations of bacteria and other microorganisms, which help keep it fit for the growth of plants. Since all these living things are concentrated in the upper layers, where roots grow, to allow fourteen feet for the subterranean depth of the biosphere should be adequate. The few bacteria or other microorganisms that may be washed deeper by percolating water are a very minor element in the biota.

This gives us 264 feet, or one-twentieth of a mile, as the average thickness of the biosphere over the land. Since the total area of land (including the vast, lifeless glaciers of Antarctica and Greenland) is approximately 57 million square miles, this gives us 2.85 million cubic miles as the volume of the terrestrial section of the planet's biosphere.

In the oceans, life extends from the surface down to the greatest depths of about 35,800 feet. With the exception of a certain amount of food carried into the seas by rivers, all marine animals depend for their nourishment on marine plants, which can carry on photosynthesis only in the sunlit upper strata, down to a depth of perhaps 600 feet in the clearest water, although for most species, in many seas, effective photosynthesis is hardly possible below 200 or 300 feet. Just as the winged creatures that fly above the terrestrial vegetation are, without exception, nourished by food manufactured below them, so those that live at greater depths in the oceans—down to the tremendous pressure, perpetual cold, and eternal night of the abyssal trenches—depend for almost all their food on plants that flourish far above them. This food reaches their levels either as dead bodies sinking downward or by long, complicated food chains that progress ever deeper.

Nevertheless, there is a great difference between the creatures that fly in sunshine above the treetops and those that swim in darkness below the limit of photosynthesis. Unless involuntarily blown away by strong air currents, the former return to their true home amid the sheltering vegetation to rest and to reproduce. Animals of the great marine depths stay there all their lives, eating, resting, and reproducing at their own levels. If they rise much above this, the air in their bodies, especially that in the swim bladders of many fishes, expands as the pressure of the surrounding water decreases. Their increased buoyancy may defeat their efforts to regain their depths, and they die horribly, torn by expanding gases.

Accordingly, the marine section of the biosphere must be taken to include the whole vertical range of the oceans. Moreover, it is coextensive with their areas, for the oceans are everywhere habitable by the creatures adapted to them; they include no barren wastes of arid sands or rocks, no lifeless glaciers—on and beneath the floating pack ice, hardy animals thrive. The mean depth of the oceans is estimated at about 12,500 feet. If we add to this 150 feet to include the many marine birds that constantly fly over it (the same allowance that we made for birds and insects flying above the treetops), the average thickness of the marine biosphere becomes 2.4 miles. The oceans, including the inland seas connected with them, cover approximately 140 million square miles. This gives us, as the volume of the marine division of the biosphere, 336 million cubic miles, which is 118 times as great as that of the terrestrial division.

Moreover, life has flourished in the seas very much longer than on land, possibly five or six times as long. Yet, by almost any criterion—anatomical, physiological, psychic, or simply number and diversity of species—terrestrial life has forged far ahead of marine life. Why? This is a question to which we shall presently return.

The Components of the Biosphere

The essential components of the biosphere are water, air, soil, and sunlight. Surprisingly, the greater part of the biosphere lies in perpetual blackness, below the depth of a thousand feet or so to which the last feeble rays of sunshine penetrate in very clear water. Nevertheless, all living things, including those in lightless caves as well as those in the oceans' depths, are at least indirectly dependent upon sunlight for the manufacture of their food by photosynthesis, the only exceptions being those bacteria that can obtain energy for carbon assimilation by the oxidation in the dark of such simple compounds as ammonia, hydrogen sulfide, nitrous acid, ferrous oxide, or methane. Nearly the whole of the marine section of the biosphere is far removed from soil; only on atolls and in the shallow, lighted waters of the continental shelves can plants grow in contact with it.

Although it is likewise true that by far the greater part of the biosphere—all that below the surface of the seas—is not directly in contact with the atmosphere, nearly all the organisms that dwell in the depths are dependent upon dissolved atmospheric gases: oxygen for respiration, carbon dioxide for photosynthesis. These gases enter the seas at the surface, a process accelerated when breaking waves mix air with water, and are borne downward into the abyss by descending cold currents, which, far below the surface, flow from the polar regions toward the equator. Since the solubility of oxygen in water decreases with rising temperatures, warm seas tend to be more sparsely populated with living things than cold seas, although the former support more species.

When Shakespeare made Hamlet call the air "this most excellent canopy," he could not have known all its claims to excellence. Since the atmosphere's composition was not discovered until many years later, he was not aware how its various gases complement each other, forming a mixture peculiarly favorable for life. All plants and animals, except a number of anaerobic microorganisms, need oxygen for respiration; but the 21 percent of this gas in the atmosphere is quite enough for them, as it is for the fires that cook our food, heat our homes, and are indispensable in metallurgy and innumerable industrial processes. Adaptable living organisms could doubtless with time adjust their metabolisms to great increases or decreases in oxygen; but in an atmosphere composed largely or wholly of oxygen conflagrations of all kinds would be more frequent and disastrous, and cities and forests would be more often destroyed. Accordingly, we should be grateful that the air we breathe contains 78 percent of the much less active nitrogen, while the remaining 1 percent is made up chiefly of argon and minute amounts of the other "noble" or chemically inert gases—neon, krypton, helium, and xenon—along with .03 percent of carbon dioxide and various impurities. A most important constituent of the atmosphere is its water vapor, which, unlike all the other gases, liquefies at the usual atmospheric temperatures, so that its amount varies greatly from place to place and hour to hour.

The atmospheric oxygen has another function, second only, in its importance to life, to that of supporting respiration. Above the highest clouds, at altitudes between about twenty-five and forty miles, a proportion of the oxygen atoms combine in trios, known as ozone, instead of the usual doublets of gaseous oxygen. The total amount of ozone is extremely small, being the equivalent of a layer surrounding Earth, at atmospheric pressure and room temperature, only one-tenth of an inch thick, yet it so effectively absorbs the Sun's ultraviolet radiation that only a small fraction of it reaches Earth's surface. These short waves of high energy injure living tissues, including the eyes; but we need some of them to form vitamin D, which prevents rickets. Too much ozone would keep this essential minimum from reaching us. Life has had to adjust itself to the physical conditions it encountered on Earth; but, if these conditions had been in any way too extreme, it might have failed in this delicate task. It is fortunate that most of the planet's oxygen is chemically bound; it accounts for about 47 percent by weight of Earth's crust and 86 percent of the oceans.

Of the four basic components of the biosphere, soil is the most variable in physical and chemical properties. At one extreme we have coarse sands, highly permeable to water and air, of slight water-holding capacity, and poor in nutrients for plants. At the opposite extreme are fine, dense clays, unfavorable for vegetation because of their impermeability. The best agricultural soils are friable loams, a mixture of coarse and fine particles, including much colloidal material, rich in the minerals that plants need, with a liberal admixture of organic remains in the form of humus and a considerable capacity to retain water.

Except for humus, all the soil's constituents are derived from the weathering of the crustal rocks under the action of rain, wind, frost, and sunshine. Residual soils, formed where we find them by the disintegration of the underlying rocks, can become very deep; but, especially in regions of high rainfall, leaching removes much of their mineral nutrients, leaving them rather sterile. Among the richest soils are alluvium, deposited in valley bottoms and on the floodplains of rivers, and loess, accumulations of fine sand and dust blown by strong winds from deserts or high, dry plateaus. Although lava flows sterilize the land, volcanic ash often fertilizes it; the slopes of old volcanoes may support thriving farms and gardens.

Of all the adaptations of plants, that to the chemical composition of the soil is among the most important but the least obvious, so that we may try in vain to grow certain introduced species on land where the native vegetation thrives. Soil, the legacy of dying rocks, is the most fragile and easily exhausted item in our natural patrimony. We may prolong its productive life with chemical fertilizers; but, unless we husband it carefully, it will eventually lose its vitality or be eroded away, leaving us bankrupt.

Many animals and even plants, including floating aquatics and the multitudinous epiphytes and parasites that burden trees in the tropics, never touch the soil, and some organisms can grow and reproduce without light or without air; but all active life is strictly dependent upon water. Nevertheless, many animals never drink and possibly never even touch liquid water. Insects that live in grain, wood, or paper may find a little moisture in these apparently dry materials, but they depend chiefly upon metabolic water or water derived from the oxidation, in their tissues, of the carbohydrates or other foods that they eat. Likewise, metabolic water suffices for some small rodents, birds, insects, and other animals of arid deserts, who may pass months, possibly their whole lives, without drinking.

The Uniqueness of Water

This liquid, indispensable for life, is in many ways a unique substance. The air we breathe is a mixture of gases. We are all familiar with a variety of solids—soils, stones of different kinds, crystals, wood—even apart from those made by man. But we may live a long while, and travel far, without meeting more than a trace of any natural liquid except water. For long ages, before they invented even such a simple art as pressing oil from olives or other vegetable productions, the great majority of our remote ancestors probably never knew any other liquid. Of course, they were familiar with the juices of plants, and blood, and doubtless the nectar of flowers, but these are all water with certain additions. In fermenting fruits, they probably tasted a little dilute alcohol, but man never knew pure or even highly concentrated alcohol until he invented distillation.

Here and there, a wanderer may have noticed petroleum oozing from the ground. Even more rarely, he may have discovered a few drops of native mercury and been fascinated by its glitter and elusiveness. But, nearly everywhere, water was the one liquid that Paleolithic man was likely to know, so completely had it taken possession of Earth's surface, to the virtual exclusion of all others. Not only does it cover 71 percent of the planet's surface and periodically deluge much of the rest, but it penetrates the crust to great depths, steams out through volcanoes and geysers, moistens the air, floats high in the atmosphere as clouds, and enters intimately into the composition of many crystalline minerals. Life has become utterly dependent upon it.

A revolving planet that exposes its face alternately to an incandescent star and the great cold of outer space, one that is, moreover, tilted toward the plane of its orbit around the luminary, is inevitably subject to great changes in temperature. Could such a planet have chosen, of all known substances, that best able to mitigate these changes, it could not have made a better choice than water. Chemically, water is compounded of two active gases so strongly bound together that they separate spontaneously only at temperatures far higher than any ordinarily found on Earth's surface. It is one of the most versatile of solvents, but it is exceptionally resistant to physical changes.

To raise the temperature of water one degree requires thirty times as much heat as must be added to raise an equal weight of mercury one degree, nine times as much as for iron, five times as much as for granite or marble, two and a half times as much as for wood, and nearly twice as much as for ethyl alcohol. Accordingly, as the temperature rises in spring and summer, a large body of water absorbs much heat and moderates its effect on the neighboring land. Similarly, as the temperature falls in autumn and winter, the water releases much heat into the air and retards its chilling. For this reason, islands and continents in the path of steady winds from the sea enjoy a mild maritime climate, while great expanses of land at high latitudes suffer from a harsher continental climate.

Water's great heat capacity makes oceans currents highly effective in transporting tropical warmth to latitudes that receive much less solar radiation. Without the Gulf Stream, northwestern Europe, instead of being a region of great population, culture, and industry, might be as bleak and barren as Labrador on the opposite side of the Atlantic. Flowing through the Strait of Florida with a volume several hundred times as great as that of the Mississippi River and a velocity of three miles per hour, this current carries a vast amount of warm tropical water that, after crossing the North Atlantic, it distributes along the western coasts of Europe, from Spain to Spitzbergen, to the great benefit of millions of people.

As it freezes at 32 degrees Fahrenheit, fresh water relinquishes as many calories as are needed to raise the same amount 144 degrees. The congelation of five quarts of water, already at the freezing point, releases sufficient thermal energy to raise four quarts of water from the freezing to the boiling point. As large bodies of water become covered with ice, the transfer of so much heat to the air above them retards the cooling of the latter. As they cool, liquids, like solids, commonly contract in volume; and the continuation of this shrinkage during congelation makes the resulting solid heavier than the liquid that formed it, so that it sinks to the bottom. Cooling water behaves differently. At about 7 degrees Fahrenheit above the freezing point, it stops contracting and begins to expand, with the result that the ice floats, covering deep ponds and lakes with a protective coat which prevents their solidification to the bottom and permits fishes and other organisms to remain active beneath the frozen surface. Seawater behaves similarly, but its freezing point is substantially lower than that of fresh water. Although exceptional in becoming lighter as it solidifies, water is not altogether unique. Solid pieces of iron and bismuth float upon the molten metals, but at temperatures far too high for life. If water behaved more conventionally and sank as it froze, oceans might become solid masses of ice, making Earth scarcely habitable.

In yet another way, water resists change. Salts, sugars, and other dissolved substances lower the freezing point of a liquid and elevate its boiling point. The alteration of these points for a given amount of the solute is much less for water than for other liquids. Seawater of average salinity freezes at 28.6 degrees Fahrenheit, which is 3.4 degrees below the freezing point of pure water. If water were as strongly affected by solutes as are many other liquids, the depression of the freezing point would be several times as great, and arctic seas, as well as the ocean's depths to which polar water sinks, would be correspondingly colder. The climate at high latitudes would also be more severe.

Water appears to be stubbornly determined to preserve the liquid state in which life needs it and the moderate temperatures that permit vital activities. It resists evaporation even more strongly than it resists freezing. The heat needed to evaporate one quart of water at room temperature would raise nearly six quarts from the freezing to the boiling point in a closed vessel that prevents loss of vapor. At the boiling point at sea level, the latent heat of vaporization is only slightly less. To evaporate most other familiar liquids, very much less heat is needed: for ammonia, about half as much; for ethyl alcohol, considerably less than half as much. Just as freezing water helps prevent the air becoming colder, so, on warm days, evaporating water cools the air. Water's high latent heat of vaporization makes sweating or panting an efficient way to cool an overheated animal body and transpiration an effective means of preventing foliage from becoming too hot in sunshine.

As a coating of ice retards the refrigeration of deep water, so the variable amount of water always present in the atmosphere, as clouds or invisible vapor, forms a protective blanket over the planet. Water vapor is highly permeable to the visible light in which the Sun sends most energy to Earth. Absorbed by soil, rocks, vegetation, or whatever else it strikes, this radiant energy raises their temperature and causes them to radiate heat more strongly. Being very much cooler than the Sun, these terrestrial bodies release their energy in waves longer than those that brought it to them, in the infrared region of the spectrum rather than the visible region. The atmospheric moisture absorbs these longer waves much more strongly than it does the visible light from the Sun, and part of the heat so retained is radiated back to the ground instead of escaping into outer space.

The water vapor in the air acts like a greenhouse, which even without internal heating is warmer on a clear, cold day than the outside air, because the glass permits the sunlight to enter freely but is rather impervious to the longer waves emitted by the warmed plants and other things inside. Clouds intercept much more of the solar radiation than invisible water vapor does, but they even more effectively prevent the escape of heat from the ground. In tropical highlands, frost rarely forms beneath an overcast nocturnal sky, even at seasons when the dawn following a starry, windless night consistently reveals fields white with hoarfrost.

The atmospheric water vapor complements in a most important way the oceans' role in stabilizing climate. A great reduction in its quantity, permitting Earth to radiate into space much more of the heat it receives from the Sun, might cause glacial conditions over much of the planet. Indeed, one of the many conflicting theories of the origin of the Ice Age attributes it to just this cause. Much water vapor remains in the upper atmosphere because it lacks nuclei on which to condense in drops large enough to fall as rain. A continuous and bountiful supply of such nuclei would increase precipitation and keep the atmosphere less humid. The nuclei might be provided by the minute meteors, much smaller than visible shooting stars, that Earth sweeps up as it speeds through space. If, for long intervals, the planet continued to traverse unusually dense clouds of this meteoric dust, the atmosphere might remain so constantly poor in water vapor that great ice sheets would cover much of the continents.

Without enough atmospheric water vapor to retain much of the radiation from Earth while it permits the inward passage of solar radiation, all water's other helpful attributes—its thermal stability and great capacity for heat that it distributes by means of ocean currents, its resistance to freezing and capacity to cool by evaporation—all these together might prove inadequate to keep the biosphere highly favorable for life.

Water's stubborn resistance to change has its reverse side. If, for any reason, the climate deteriorates to the point where immensely thick sheets of ice cover the continents, recovery is slow. For, just as water relinquishes much heat as it freezes, so must an equal amount of heat be returned to the ice in order to melt it. Moreover, the reflection of sunshine from the glittering surface of snow and ice retards absorption of this heat. If ice liquefied as readily as many other substances at their melting points, recovery from a glacial period would be more rapid. Nevertheless, the slow shrinkage of glaciers as the climate improves (as was happening during the first half of this century) is not without advantages, as it gives the living world time to adjust to great changes. If the immense masses of ice, remnants of the last glacial period, that now cover Antarctica and most of Greenland were suddenly to melt, sea level everywhere would rise several hundred feet, flooding great areas of lowlands, including some of the world's major cities.

Another attribute in which water surpasses most other liquids is its cohesion. Although the molecules of a liquid slide over each other easily, making it fluid, they resist separation. This cohesion is the source of surface tension, which causes water to rise in a capillary tube that it wets and mercury to sink in a glass tube that it refuses to wet. The top of the water column has a concave meniscus that pulls it upward above the level of the surrounding liquid; the meniscus of the mercury column is convex, depressing it. The surface tension of water enables us to improvise a compass by floating on its surface a magnetized sewing needle thinly coated with oil, as by rubbing our fingers over it, to make it unwettable. The liquid behaves as though it were covered by a very thin skin of stretched rubber, easily broken by any intruding solid, but immediately repairing itself when the foreign object is removed. At room temperature, water in contact with air has a surface tension more than four times as great as that of ether, more than three times that of alcohol, more than twice that of benzol and olive oil. The surface tension of mercury, however, is seven times that of water. Iron, copper, silver, gold, and platinum have a much greater surface tension than water, at the far higher temperatures at which they melt.

Surface tension and allied phenomena resulting from the cohesion of water have important consequences for living things. Surface tension enables water striders to rest and skate around upon the water, which they barely dimple with four of their six feet, tufted with unwettable hairs. Whirligig beetles skim rapidly over the water's surface. The larvae of mosquitoes and nymphs of water scorpions hang suspended from the surface film. For a greater number of insects, however, surface tension can have unfortunate consequences. When thoroughly wet, a small insect, such as a fly, is impeded in its movements not only by a load of clinging water that may weigh several times as much as it does but also by the surface tension that tends to bind its wings and legs together. It may drown, or be snatched up by some insectivorous animal, before it can dry.

In appropriate conditions, the cohesion of water gives it high tensile strength. We might never suspect this as we watch water drip in a discontinuous stream from a leaky faucet, but it can be demonstrated by ingenious experiments, and it enables trees to grow much higher than root pressure could force water upward. This pressure, which causes sap to flow from the trunks of sugar maples in early spring, before the leaves have expanded, is insignificant or absent in full-foliaged trees transpiring rapidly in sunshine. In them, water is not pushed upward from below but pulled up from above.

The sap in the narrow vessels of wood behaves like so many thin wires being hauled up by forces in the leaves.

The seat of these forces appears to be the submicroscopic concave menisci in the cellulose walls adjoining the intercellular spaces of the leaves, which lose water as it diffuses outward through the open stomata of foliage engaged in active photosynthesis. The pull exerted by innumerable concave surfaces, whose surface tension tends to decrease their curvature, is transmitted through the turgid leaf cells to the water columns in the wood vessels and keeps the liquid flowing upward, in tall trees very much higher than it could be forced by atmospheric pressure. For these wire-like water columns to continue to move upward, they must remain uninterrupted and in unbroken contact with the supporting walls of the vessels. If an air bubble develops in a vessel, it is, at least temporarily, excluded from the conducting system. But a massive trunk contains many vessels; and each year, as the tree puts forth fresh foliage, it forms a new ring of them to replace older ones that may have become filled with air or otherwise clogged.

Ferns have found a very different use for water's cohesion. Their spores are produced in dots or lines of clustered capsules, on the undersides of their fronds or sometimes along the margins. Each little spore capsule or sporangium is embraced by its annulus, a belt of special cells with thick inner and side walls and thin outer walls. As the ripe capsule dries, the contraction of the water in these cells pulls their side walls together and the outer wall inward, thereby reducing the curvature of the annulus and tearing the spore case open. When finally the tension of these cells reaches the breaking point, the annulus snaps forward like a released spring, throwing the minute spores into the air. An ingenious German botanist, O. Renner, used the germ's spore-dispersal mechanism to measure the tensile strength of water. He demonstrated a negative pressure of about 350 atmospheres, which is sufficient to pull up a column of water 2 miles high. Theoretically, the tensile strength of water is much greater than this, enough to pull it to the top of a tree 40,000 feet high!

Why Has Terrestrial Life Shot Ahead of Marine Life?

Now that we have surveyed, rather summarily, the biosphere and its varied resources, let us return to a question that we raised earlier in this chapter: why has life advanced so much farther on land than in the oceans, which offer well over a hundred times as much living space and have supported organisms possibly five or six times as long? Whether we survey the vegetable or the animal kingdom, we find the most highly evolved forms on land or, at least, of terrestrial origin. The seed-bearing plants, especially the angiosperms or flowering plants, are universally recognized to be the most advanced members of the vegetable kingdom, and all grow on land, except a small minority—all monocotyledons derived from terrestrial ancestors—that have invaded the margins of the seas. Moreover, the terrestrial flora is incomparably richer than the marine flora, with an estimated 250,000 species of angiosperms, over 20,000 kinds of mosses and liverworts, 10,000 of ferns and their allies, and 700 of conifers and their allies, as opposed to only about 20,000 species of algae, including those of fresh water.

Of vertebrates, the most highly evolved group of animals, about equal numbers of species inhabit the land and the water. Homeothermy, or the ability to maintain a nearly constant body temperature despite large fluctuations in the ambient temperature—one of the last refinements of animal physiology—is found only in birds and mammals, essentially terrestrial creatures, although a few, such as the whales and dolphins, have from terrestrial ancestors become wholly marine. Many birds derive most or all of their food from the water, but all breathe air, and even the most aquatic of them nest above rather than in water. Finally, terrestrial animals exploit the oceans as marine creatures cannot exploit the land, and—a less debatable criterion of evolutionary advance—certain terrestrial animals explore the seas and try to learn their secrets, although no marine animal appears to be curious about the continents.

When we compare the marine and terrestrial environments, the greater progress of terrestrial life appears the more inexplicable. Marine organisms pass their lives in the unique liquid of which their bodies are chiefly composed. It contains in solution all the essential minerals and, in most oceanic waters, adequate amounts of oxygen. The upper layers of the oceans receive sufficient sunlight, and contain enough carbon dioxide, for highly productive photosynthesis. Contrast the simplicity of life in the sea with the problems that confront terrestrial organisms! All, plants and animals alike, must somehow conserve indispensable moisture in an atmosphere that is frequently drying. Green plants must extract it from soil that is often nearly waterless and transport it to the aerial organs where it is used for photosynthesis. All the larger of them need more diverse organs and cells, greater anatomical differentiation, than do seaweeds that can absorb water and nutrient salts through practically their whole surface. For sexual reproduction, terrestrial plants cannot simply discharge their sexual cells into the water so that they may swim or float together; they must develop less simple methods of insuring fertilization. In many terrestrial habitats, plants and animals alike confront greater fluctuations in temperature, availability of water, and other conditions than ever affect marine organisms. It is easy to understand why it took life such long aeons to conquer the land. But why, after it had become well established there, did it progress so much more rapidly than the life that remained in the seas?

Apparently, it was just the difficulties that terrestrial life confronted, and had to overcome, that promoted its advance. Marine life could idle in its well-tried ancient ways; its physiological problems had long since been solved; its anatomical structures had been proved by ages of service. Terrestrial life had to find new solutions or perish. Just as man has become what he is by being confronted with, and solving, many difficult problems; and each of us, individually, becomes more mature and capable by resolutely meeting and overcoming challenges; so terrestrial life has profited by the very obstacles that it encountered.

Not that prehuman creatures rationalized their problems and worked out solutions in their minds. Far from it! They had to await slow evolutionary changes, which depended upon random mutations and their testing and screening by the total environment, physical and biological. But, in many cases, the result was much the same as though they had deliberately tried to solve their problems and had, moreover, the power—which we lack—of building the solutions into their tissues and vital functions. This gave terrestrial organisms great diversity, advanced anatomical structure, adaptability, and the hardihood to withstand environmental extremes, and terrestrial animals acquired a general level of intelligence above that of marine animals. It is no accident that the most intelligent of all animals inhabits the land rather than the water and that the cetaceans, derived from the land, are becoming recognized as the most mentally advanced inhabitants of the oceans.

Other factors, too, have helped terrestrial animals advance in social organization and intelligence. Not the least of these is the possibility of having a territory or a permanent abode. Except on reefs of coral or rock and in the shallow waters of the continental shelves, and doubtfully on the bottoms of deep oceans of which we know too little, marine animals can have no fixed habitation; the millions of cubic miles of oceanic waters lack landmarks to distinguish one locality from another. The wanderings of these homeless pelagic animals appear to be limited only by differences in such factors as the temperature, salinity, and pressure of the water. In these circumstances, an integrated social life is difficult to achieve (although the cetaceans preserve a measure of it), and parental care, the nursery of altruism and personal attachments, is only rarely feasible. The absence of a stable substratum, as of suitable materials, prevents the construction of nests—the centers of the social life of many insects and birds. The arts and crafts that have exercised and strengthened man's practical intelligence—ceramics, weaving, metallurgy, building, agriculture, writing—could hardly flourish underwater.

Savage as the life of the continents can be, that of the oceans is far more savage, more relentlessly predatory. The larger terrestrial mammals, from antelopes to elephants, are mostly vegetarian grazers and browsers. Many smaller terrestrial animals exchange benefits with plants, pollinating their flowers in return for nectar, scattering their seeds in payment for succulent fruits. A few, including man and certain ants and termites, have developed agriculture, cultivating the plants that nourish them. In these several ways, the strife of nature is mitigated. The high seas support neither flowers nor seed-bearing fruits. On the reefs and continental shelves, certain fishes and other animals graze or browse on seaweeds, and the algae floating in the vast Sargasso Sea in the middle of the Atlantic offer opportunities for a vegetable diet.

Elsewhere on the high seas, the principal agents of photosynthesis are the microscopic plants of the plankton. Here the chief producers of food are diatoms, brownish algae enclosed in delicately sculptured siliceous boxes of the most varied shapes, of which millions may float in a quart of sunlit water. Next come the dinoflagellates or peridinians, some of which contain chlorophyll and synthesize their own food, while others depend upon living or dead organic matter and are sometimes classified as animals. One form abundant on the high seas is a single minute cell with three hornlike projections that help it keep afloat and two cilia that propel it through the water. These and other planktonic plants yield food that in nutritive value and total weight compares favorably with the product of the terrestrial vegetation composed of larger plants.

This grass of the oceanic meadows is too minute to be grazed by large animals, like the herbage of prairies and savannas. It is consumed in vast quantities by small planktonic animals, chiefly such crustaceans as the tiny copepods and the slightly bigger euphausid shrimps, which in turn are devoured by larger animals—and so successively, until the great sharks and octopuses and whales prey upon creatures of fair size or upon each other. Somewhat exceptional are the huge whalebone whales, the almost equally big basking sharks and whale sharks, and certain middle-size fishes such as the mackerel and the shad, all of which shorten the food chain by sifting from the water immense numbers of the small crustaceans, often known collectively as krill, that collect such primary food producers as the diatoms and dinoflagellates. In the pitch-black depths of the oceans, by far the greater part of their vertical sweep, every animal that is not a scavenger is necessarily a carnivore, for no plants synthesize food there.[1]

The incongruously huge mouths of many of these marine animals, their formidable arrays of teeth, remind us vividly of the unrelenting fierceness of predation in the oceans, where no friendly vegetation shelters and conceals the hunted, no nook or burrow in rock or soil offers a safe retreat. Although, physiologically, life is easier in the seas than on land, the absence of shelter from teeming predators makes it much more perilous. Marine animals must eat other animals in such vast quantities because in the aggregate they so greatly outweigh the plants that support them. The total mass of animal life in the oceans has been estimated to be about six times as great as that of vegetable life. Only by passing their substance from one to another can such an overwhelming preponderance of animals remain alive. With each step in a food chain, some energy is lost, and this maelstrom of eating and being eaten would soon exhaust itself if

[1] Since this was written, explorations of the oceans' floors at great depths have revealed fissures through which hot water rich in dissolved minerals wells up into the cold seawater. Like oases in the desert, these hot springs support a varied assemblage of bizarre invertebrate creatures, all ultimately dependent upon bacteria that oxidize hydrogen sulfide for their energy. Nevertheless, it remains true that most life in the oceans' depths is nourished by photosynthesis in the sunlit layers far above.

the primary food producers that keep it going, chiefly the prodigiously abundant diatoms, did not multiply so rapidly.

In contrast to the situation in the oceans, the terrestrial flora has been estimated to weigh more than one hundred times as much as all the terrestrial animals. Although much of this mass of vegetation is wood that only termites and a few other insects can eat, whereas the microscopic plants of the sea are swallowed whole by copepods and other small creatures, the terrestrial flora nevertheless enables a much larger segment of the fauna to live without eating other animals.

Nothing more clearly emphasizes the perils of life in the oceans than the number of eggs that its inhabitants must produce to maintain their populations. A single cod may lay as many as 9 million eggs; an oyster may release 500 million in a single season. Among terrestrial animals, even the prolific insects cannot compete with such fecundity. A queen termite, reproducing for a large colony, is credited with a million or so eggs in her lifetime. Otherwise, few insects appear to lay more than two or three thousand, while one or two hundred is more usual. No terrestrial vertebrate remotely approaches the fecundity of marine fishes; many birds and mammals produce no more than one, two, or three eggs or young in a year.

The seas gave birth to life and, as great reservoirs of water and heat, make the continents more habitable. But the land, despite all the problems that it presents to living things—and in large measure because of these problems— is where life flowers, in the figurative no less than the literal sense. To advance their psychic life, animals needed a less dense medium that permitted wider visibility, a more richly varied vegetation to provide food and shelter, the possibility of a fixed abode where families could remain intact, a more diversified habitat that made greater demands on their adaptability and intelligence, and some mitigation of predation. Perhaps, too, they had to move a little closer to the stars.

4. Plants, Sustainers of Life

The fortunate planet differs strikingly from the eight sisters that circulate with it around the Sun. It does not expose a bleak, naked surface, nearly or quite waterless and lifeless, like Mercury and Mars. Nor does it hide its face beneath a dense, heavy atmosphere, like Venus and the massive outer planets. Its air is transparent (when not polluted by man) and rich in oxygen, as on no other planet, as far as we can learn. Innumerable animals of many kinds fly over it, walk on its surface, or swim in its water, as nowhere else in the solar system. Finally, its lands, where not too cold or dry or denuded by man, are covered with green vegetation, a living garment found only here. And this verdant mantle explains most of the other differences that we have noticed.

Astronomers believe that the young planet was too hot to hold an atmosphere; whatever gaseous molecules surrounded it darted with such great velocity that they shot off into space and were lost. As the more superficial rocks cooled, gases, including water vapor, escaped from them; and now, less strongly agitated by heat, they were held by the planet's gravitational field. Be this as it may, there is widespread agreement that, before life arose, Earth's atmosphere was very different from that which we know. It consisted largely of carbon dioxide, as on Venus and Mars, with considerable amounts of methane, ammonia, and probably some free nitrogen and free hydrogen, but scarcely any free oxygen. Likewise, it contained a vast amount of water vapor, most of which condensed with further cooling to fill the oceans.

Experiments have shown that amino acids, the building blocks of proteins, accumulate when an electric current is passed through a mixture of gases such as the primitive atmosphere contained. They may also be formed in such a mixture by the momentary great rise in temperature that accompanies the passage of a shock wave. The frequent thunderstorms that doubtless shook the turbulent early atmosphere supplied both of these conditions, while falling meteorites produced the latter, thereby generating amino acids that drifted down into the primeval seas or were washed into them by rain. The composition of these bodies of water differed greatly from that of today's oceans. They were warmer, with less of the salts that, over the ages, rivers have delivered to them from weathering rocks; but they were richer in amino acids, hydrocarbons, and other substances that living things could use; they have been compared to a hot, thin soup. In these early seas or, more probably, in coastal lagoons where their rich content was concentrated by evaporation, life somehow arose, probably as long as 3,000 million years ago.

The eobionts, as the tiny, primitive forerunners of life are called, seemingly possessed two properties that have been indispensable to all their descendants: they enclosed themselves in some sort of pellicle or membrane that regulated their exchanges with the surrounding water, and they could multiply themselves. Still unable to synthesize their own food from simple inorganic substances, they depended for nourishment upon elaborated materials, probably chiefly amino acids formed in the atmosphere above them. Evidently they could live with little or no oxygen, like many anaerobic bacteria of the present day. Probably, before long, the eobionts developed a trait conspicuous in contemporary organisms: multiplication up to the limit of the environment's capacity to support them. The nutrients in their natal waters, which had obviously been accumulating for long ages before life arose, could not continue indefinitely to meet their increasing demands. Except possibly for traces too minute to be detected, such nutrients have long since vanished from our seas. Before any considerable amount could accumulate today, it would be exploited by the teeming microscopic life of the waters.

The early promise of abundant life for a privileged planet might never have been fulfilled without a second development, a step comparable in magnitude to the genesis of life itself, certainly no less important—for all subsequent life on Earth, with a few minor exceptions, would depend on it. One or more of the eobionts developed a complicated process for capturing the energy in sunlight and storing it in a compound made from two of the simplest, most abundant substances then present on the planet's surface: water and carbon dioxide.

These saviors of the living world were evidently green with chlorophyll, the pigment that colors vast areas of Earth's surface today and is the active agent in photosynthesis even in foliage that is otherwise tinted, as in many ornamental plants, and in seaweeds that are red or brown. Probably they were blue-green algae or something very like them. In any case, forms like those of blue-green algae have left their impressions on some of the earliest fossil-bearing rocks, and the extant members of this group are among the simplest autotrophic or self-supporting plants. Today, they are exceedingly humble members of the vegetable community, forming dark green, slippery patches on wet, shaded soil or rocks, growing over submerged plants, or floating in still ponds or seas. Sometimes they are violet or red, with a pigment supplementary to their chlorophyll, and so abundant that they tinge the water where they float, as in the Red Sea. Their tiny cells, lacking a definite nucleus, grow in threadlike chains or small, gelatinous globules. Some of the filamentous forms, particularly species of the well-named genus *Oscillatoria*, have the surprising habit of sliding back and forth and swaying from side to side, frequently reversing direction and, under the microscope, appearing to move quite rapidly.

Blue-greens readily enter into symbiosis with other plants of the most diverse kinds. Closely embedded in a maze of fungal strands, they form lichens, many of which grow in places more exposed and dry than these algae could tolerate alone. They form colonies in intercellular spaces of certain liverworts, in the roots of cycads, and even in such flowering plants as the tiny, floating duckweeds and the huge-leaved *Gunnera*. It has been suggested that, carrying their symbiotic tendencies farther, they entered into a still more intimate union with the cells of other plants, finally becoming the chloroplasts almost universally distributed through the vegetable kingdom, with the exception of fungi and other parasites and saprophytes. Although these algae of very ancient lineage are so small and inconspicuous today, the trees that dominate the forests and the herbs that cover the meadows may be descended from pioneer blue-green algae that discovered how to capture sunlight and thereby set the whole living community on its long upward way.

The advent of photosynthesis transformed the world—not only the living world but the face of the planet and the atmosphere that surrounded it. No longer dependent upon a diminishing supply of nutrients slowly formed by inorganic processes, organisms could become larger, more numerous, and more varied. Hitherto, it would have been difficult to say whether the eobionts were plants or animals; now, the simple creatures that made their own food were definitely becoming plants. Before long, the organisms that failed to develop this independence began to eat the green ones, thereby becoming the forerunners of the animal kingdom. Possibly even before photosynthesis arose, some of the eobionts preyed upon others, as today some of the smallest organisms do. In any case, as the living world, nourished by the products of photosynthesis, became crowded and competition for food grew keen, animals started to devour each other as well as plants. The long, cruel history of predation began while life was young.

In photosynthesis, energy derived from sunlight is used

to combine carbon dioxide with water to form a sugar. In this process, one molecule of oxygen is released for each molecule of carbon dioxide that is used. As autotrophic plants increased in abundance, more and more of the carbon dioxide in the atmosphere or dissolved in the water was replaced by oxygen. The most direct way to release the energy stored in the sugars, starches, and their derivatives that are products of photosynthesis is to reverse the process and oxidize them. Accordingly, the living world, which appears originally to have been anaerobic, turned increasingly to the respiration of oxygen, with the production of carbon dioxide.

If this reversal of the process had been complete, it would have prevented any considerable increase in the supply of free oxygen. However, organisms that took advantage of the new sources of food to grow larger and more complex did not wholly disintegrate when they died. Many marine creatures enclosed themselves in calcareous shells which, accumulating on the bottoms of the seas, created immensely thick deposits of limestone, marble, and chalk, forms of calcium carbonate that hold vast quantities of carbon which once circulated in the atmosphere and the oceans as carbon dioxide. Another great storehouse of carbon is petroleum, a product of the partial decomposition of dead organisms buried beneath layers of mud or sand that turned to rock, although an alternative origin has been suggested. After vegetation grew heavily over the land, some of its remains became coal or peat, imprisoning additional quantities of carbon. And, in every epoch, a not inconsiderable amount is locked up in living organisms of land and seas.

So much carbon dioxide has been removed from the atmosphere since plants began to consume it in photosynthesis that it now accounts for only .03 percent of the constituent gases, while the originally scarce free oxygen has increased to 21 percent. If all fossil fuels were burned to reduce limestone and chalk to quicklime, and all organisms were to die and decompose, carbon dioxide might again replace oxygen in the atmosphere. The recent great consumption of fossil fuels, along with the cutting and burning of tropical forests, is causing a slight increase of atmospheric carbon dioxide, which eventually may strongly affect Earth's climate.

The increase of free oxygen caused the virtual disappearance of the methane, ammonia, and other oxidizable compounds present in Earth's primitive atmosphere. The methane was converted into carbon dioxide and water, the ammonia into nitrogen and water, a process accelerated by lightning discharges. The removal of these gases—poisonous to recent forms of life—made the air safe for the plants and animals that presently began to venture forth from the waters.

Like a retarded infant, life lingered in its watery cradle for many millions of years before it grew strong enough to creep out upon the land. Probably at first like a toddling child that never wanders far from its mother, it braved the air in the intertidal zone of the seashore. As it cautiously advanced toward the high-tide mark, its hours of exposure to drying sunshine and winds were prolonged, but twice each day the rising sea flowed over it, replacing lost moisture and renewing its supply of mineral nutrients. Finally, the hardiest organisms severed all contact with the nourishing sea as they advanced inland over bleak, forbidding continents or islands, probably all naked rock and sand, upon which sunshine beat, winds blew, and rain pelted, untempered by a single tree or blade of grass. In this advance, the plants must have preceded the animals that depended upon them for food and shelter.

The development of a land flora took a long age because of the immense anatomical and physiological changes it demanded. The submerged alga can absorb water, salts, and dissolved gases through its whole surface. If it does not float free, it needs only holdfasts to anchor it to the rocky or sandy bottom; absorbing roots are present only in those submerged aquatics, chiefly flowering plants, whose ancestors once lived on land. Mechanical stiffening of the stem would be not only superfluous but dangerous—what tree could stand erect against the pounding of a stormy sea? Not to resist the waves but to yield to them is the strategy of the seaweed; the largest kelps, hundreds of feet long, are as flexible as ropes.

The terrestrial plant confronts a multitude of problems unknown to the seaweed. To attain any size, it must diversify and specialize its tissues to a degree never found in the water. One part must be underground to absorb water and salts, another in the air to absorb sunlight and gases. To transport water and solutes from the roots to the leaves, as well as elaborated food from the leaves to the roots, a two-way conducting system is indispensable. To hold the stem erect, mechanical elements are required. To prevent desiccation, the plant's aerial parts need an impermeable covering, a waxy cuticle or bark; yet this must not be too tightly closed, for the plant must transpire water to move its salts upward, admit carbon dioxide for photosynthesis, permit the oxygen released by this process to escape, and perhaps absorb fresh oxygen at night for respiration. Moreover, for the fertilization involved in sexual reproduction, it must develop methods totally different from those used by submerged aquatics.

Life is resourceful and, given enough time, seems able to overcome almost any difficulty that confronts it, no matter how impossible this may appear. By the Devonian period, nearly 400 million years ago, some of the basic problems of land plants had been solved so well that forests appeared. They continued to spread and to prosper until, in the following geological period, the Mississippian or early Carboniferous, they covered large areas of the northern hemisphere.

To wander through those forests that flourished over 300 million years ago, what an experience it would be! In stiff dignity, lepidodendrons tower above us to heights of over a hundred feet. Their thick trunks divide into dichotomous branches, whose younger portions bear simple leaves in spirals. The older parts of the branches and trunks are thickly covered with diamond-shaped cushions left by fallen foliage. Instead of flowers, the branches bear cones at their tips, calling to mind certain of the humble modern relatives of these ancient giants, the lycopods or club mosses. Among the lepidodendrons stand equally tall sigillarias, with narrow leaves that create longitudinal rows of hexagonal scars when they fall, and calamites that resemble gigantic horsetails. The latter bear their linear or wedge-shaped leaves in whorls and, like their modern descendants, produce cones filled with spores. Beneath these archaic trees grow ferns whose finely divided fronds have a more familiar aspect than anything else we find here. Terrestrial, arborescent, or scrambling into the lower branches of the trees, they remind us of certain ferny mountain forests of our own tropics. Some may be true ferns, but others are seed ferns that have left no modern representatives. The seeds of some, borne singly in leafy cupules sprinkled with stalked glands, are the most flowerlike objects we find here.

As we struggle over the soggy ground, littered with fallen trunks and limbs that decay slowly, we may find breathing difficult, for the air still contains less oxygen and much more carbon dioxide than that to which we are accustomed. Vast quantities of the latter will be removed from the atmosphere by these same forests, to be finally buried beneath thick alluvial deposits and converted into coal, and much more will be locked up in the great limestone formations of succeeding geologic periods. The richer supply of carbon dioxide should favor more rapid photosynthesis and growth, for this effect has been experimentally demonstrated in modern plants, but perhaps heavily clouded skies work in the opposite direction by reducing sunshine.

Even those accustomed to the paucity of bright colors in the lower levels of the heavy tropical rain forests of our modern era might find these ancient woodlands monotonously green and brown. The very conditions that favor brilliant coloration in plants and color vision in animals—flowering and fruiting plants that compete for the services both of pollinating insects and of birds and frugivorous animals that scatter seeds—have hardly yet arisen. Adding to the gloom of these dank woods is the prevailing silence. Birds of any kind are still far in the future; songbirds, farther still. Perhaps we hear the croaking of some primitive amphibian or the buzzing of some of the earliest insects. But at least we are fairly safe. No snakes lurk amid the undergrowth to sink venomous fangs into the legs of the unwary. No great cats or other large carnivores prowl amid the tree trunks, hungry for warm blood; even the ferocious dinosaurs have not yet appeared on Earth. Some big amphibians, like the three-foot-long *Ichthyostega*, have formidable arrays of teeth, but they are slow-moving fish eaters that we easily avoid. If any biting insects are about at this early age, they probably would not molest us, as they have never tasted warm blood. Perhaps our greatest peril is that of sinking above our heads into the treacherous mud.

By the Permian period, some 250 million years ago, the lepidodendrons, sigillarias, and calamites were being replaced by gymnosperms, whose modern representatives include the pines, spruces, and junipers so abundant in northern lands. Finally, in the early Cretaceous, 130 million years ago, plants of a new type were multiplying. Instead of shedding their seeds naked, they covered them, singly or many together, in some sort of envelope—a dry pod, a hard woody shell, or a fleshy fruit. Accordingly, they are known as angiosperms, a term compounded of two Greek words meaning vessel and seed. They are also called the flowering plants, although some, such as oaks and alders, are hardly more deserving of this appellation (except in a technical sense) than the pines and spruces.

The flowering plants brought into the vegetable world a greater flexibility of organization than it had hitherto known, and they proceeded to demonstrate this by rapid evolutionary changes. They could grow as herbs or shrubs or trees or vines, in the ground or perched on other plants, in arid deserts or submerged in water, as well as in moist ground, independently or as parasites. They used the most diverse agents to transfer their pollen: wind, water, insects, birds, bats, even small quadrupeds. They were equally versatile in the means that they employed for spreading their seeds far and wide. They entered into more intimate and complex relations with animals than plants had hitherto done, often in mutually beneficial partnerships. Such groups as butterflies, bees, hummingbirds, songbirds, and the primates, including man, owe their present success to the flowering plants. These animals could hardly have existed, certainly not in their present diversity and abundance, amid the older floras that covered Earth before the angiosperms arose.

Since their first appearance in the early Cretaceous, the flowering plants have spread over most of the land, gaining ascendancy nearly everywhere except in the boreal coniferous forests and in certain restricted areas of more southerly regions. About 250,000 species comprise the vast and exceedingly diverse array of angiosperms. Just as the flowering plants are dominant in the vegetable kingdom, so the vertebrates are dominant in the animal kingdom, although some might assign this position to the insects, which are much more numerous in species and individuals, except in the seas. Yet the vertebrates number only about 40,000 species, more than half of them fishes, a large proportion of which live in salt water, where flowering plants are poorly represented. The continents and islands support about 10 species of flowering plants for every species of vertebrate animals. Why this great difference? Why are there so many more kinds of terrestrial flowering plants than terrestrial vertebrate animals?

An obvious reason for the great diversity of plants involves the great diversity of habitats: wetlands and dry lands, warm lowlands and cool highlands, acid soils and alkaline soils, temperate zones and tropics, and so forth, in the most varied combinations. Moreover, plants offer habitats for other plants, especially in the tropics, where a large variety of herbs, shrubs, and even trees grow on the trunks and branches of big trees. But all these habitats are also available to animals, who do not fail to take advantage of them.

Another factor that contributes to the greater diversity of plants is their lesser mobility, compared with that of animals. To be sure, seeds of many kinds, and sometimes vegetative parts capable of starting new plants, are borne for considerable distances by wind, by flowing water, and especially by animals to which they attach themselves or which swallow them. Migrating birds are a principal agent for this wide dispersal of seeds. But many plants lack such efficient means of dispersal, and, on the whole, flowering plants are less mobile than ferns, fungi, and other cryptogams, whose lighter spores may be wafted longer distances by wind. The mobility of vertebrate animals, especially birds, makes it more difficult for local populations to preserve the reproductive isolation that enables one species to split into several species, except on islands or in isolated highland areas which are, in effect, islands in the air. The sedentary habit of plants favors the differentiation of species with restricted ranges. The species of plants are more numerous than the species of animals because they are, on the whole, more local—although a minority, even before man began to scatter them widely over Earth, achieved vast ranges.

No less important than the foregoing are certain intrinsic differences between plants and animals. Plants' competition for a place in the soil and sunlight is gentle and persistent rather than violent, and they do not devour one another. They never, like animals, expel other individuals from a territory many times their own size; but they tolerate members of the same or different species in closest proximity. Accordingly, a great diversity of plants with similar needs can thrive close together.

Because plants are less strongly individualized than animals, they have greater recuperative powers. Most flowering plants have an indefinite number of organs, such as roots, leaves, and flowers, and they can survive the loss of many of them from, for example, the attacks of insects and browsing animals; however, the loss of a limb or an eye, due to fights for mates or attempted predation, is usually soon fatal to free animals.

Plants also have greater reproductive powers than animals. The number of seeds produced by a tree, or even a vigorous shrub or herb, greatly exceeds the number of young produced by terrestrial vertebrates or even by most insects, although probably only exceptionally does it surpass the number of eggs laid by a female termite or honeybee or by certain exceedingly prolific marine fishes, such as the cod. Moreover, many plants enjoy greater reproductive flexibility than animals. In order to reproduce successfully, most animals must attain a certain size and vigor in a favorable environment. But the same plant that in rich soil grows tall and sets many seeds may in poor soil remain a depauperate specimen yet produce enough seeds to perpetuate its species. If they fail to reproduce sexually, many plants can propagate vegetatively, by runners, bulbils, detached branches, and various other means. Because plants enjoy greater reproductive flexibility than do the higher animals, unless their habitat is radically altered, as by climatic changes or by man, vegetable species often manage to persist where associated animal species vanish.

Contributing greatly to the diversity of vegetable species is the fact that natural selection is less stringent in them than in animals, especially vertebrates. The forms of vegetable organs are not so strictly tied to their functions as are those of animals. For each animal and each particular type of locomotion—walking, running, flying, or swimming—limbs of a certain definite construction are most efficient, and divergence from this form would greatly handicap the animal in the struggle for existence. A similar relation holds between teeth or bills and kinds of food, between digestive tracts and diet, between hearts and circulation, and so forth. Contrast with this the wide diversity of forms consistent with the efficient performance of the same essential function in plants. Consider the great variety of the shapes of leaves, all engaged in productive photosynthesis in the same meadow or woodland; the differences in their arrangement, alternate, opposite, or whorled; the various systems of branching of the plants that bear them. Or contemplate the immense diversity in the structures and colors of flowers that can be pollinated by the same agents—bees, butterflies, birds, or wind. And recall the many different ways, equally adequate, by which plants disseminate their seeds.

In plants, selection appears to operate more stringently upon such functions as the absorption of nutrients from the soil, photosynthesis, water storage, and transpiration than upon form, more upon physiology than upon morphology. Such selection is responsible for great differences in the tolerance of excesses or deficiencies in the elements found in the soil and in resistance to drought, cold, heat, and other environmental extremes. Although functional differences are often difficult to assess, differences in forms, recognized immediately by us, are used by botanists to separate species. This leniency of selection, this possibility to tie the same function to an immense diversity of forms, coupled with self-supporting plants' tolerance of close neighbors of the same or different species and their relatively slight mobility, are responsible for the immense variety of flowering plants that delight and fascinate us while they defeat our efforts to know them all.

(The only group of organisms with more species than the flowering plants contains the insects, which comprise about three-quarters of the estimated one million known species of the animal kingdom. There are so many species because the plants that support them are so diversified, as will be explained in chapter 6.)

The two largest families of flowering plants, the composites and the orchids, convincingly demonstrate the strongly contrasting paths that can lead flowering plants to worldwide success. The former, about 20,000 species strong, are dicotyledons, with two seedling leaves. In growth form they range from herbs like dandelions that scarcely rise above the ground, through taller herbs like goldenrod and ragweed and shrubs of all sizes, to lofty trees of tropical rain forests, and even to woody lianas that clamber to the treetops. A few species even grow as figlike epiphytes on tropical trees.

The small florets of composites grow in heads, closely surrounded by an involucre of usually green bracts. The head may be as broad and many-flowered as that of the giant garden sunflower or be reduced to a single floret, closely invested by its bracts. As a rule, pollination is not highly specialized; the same composite can often be fertilized by animals as diverse as inconspicuous insects, bright butterflies, and glittering hummingbirds. Although each fertile floret can produce only a single seed, a large floral head may yield many; since pollination is not difficult, it commonly does so. When a dandelion turns back its bracts to expose its ripe achenes, they stand close together in a plumy hemisphere, tempting children to blow upon them and watch their tiny fruits float away, each beneath its own parachute, while the few that remain attached to the flat white receptacle tell the time. Other composites, like the burdock and the bur marigolds, disperse their seeds by means of hooked appendages that stick to clothing or fur, or they enclose each seed in a berrylike fruit that attracts birds, as in the tropical American genus *Clibadium*. The composite family represents the triumph of simplicity.

Orchids belong to the smaller of the two great divisions of the angiosperms, the monocotyledons with a single seed leaf, which include the grasses, lilies, and bananas. The approximately 17,500 species of orchids range in size from miniature gems two or three inches high to long, thin, canelike growths six or seven yards tall; they are never shrubs, trees, or vines, although the long inflorescences of some species of *Oncidium* scramble upward, supported by other plants. In the tropics, the great majority of orchids grow as epiphytes on trees; but at higher latitudes, as on tropical mountains, many species are rooted in the ground. By no means all orchid flowers are as large and beautiful as those that florists sell; perhaps the majority range in size from small to minute. But all have a complex structure that contrasts strongly with the simple composite floret. The pollination of many species seems to depend upon the services of one or a few kinds of insects that are not always available or are not present in adequate numbers, with the result that, even in its native habitat, a large inflorescence of many flowers often yields only a few fertile pods or none at all.

To compensate for difficult and uncertain pollination, the orchids have developed a device that demonstrates once more the great versatility of the angiosperms. Instead of shedding their pollen in separate grains, like most plants, the orchid anthers produce pollinia, each a compact agglomeration of many pollen cells. These pollinia are attached, usually in pairs, quadruplets, or octets, to an appendage that sticks to the body of an insect. The flower that receives only one of these groups on its stigma is enabled to set a vast number of tiny, wind-borne seeds that compensate for all the neighboring flowers which fail to yield any. Thus, in the orchid family, complexity triumphs over difficulties.

It is of great interest that another family with a most complex and unusual floral structure, the dicotyledonous Asclepiadaceae or milkweeds and their relatives, also sheds pollen in pollinia, attached in pairs to a special device that facilitates carriage by insects. In milkweeds, too, pollination is rather difficult, and only a small minority of the flowers in an inflorescence may set seeds. Those that do produce single or twin pods, which split lengthwise to shed many flat seeds with silky plumes that float in the breezes—but far fewer than a typical orchid pod. Elsewhere in the vegetable kingdom, pollinia are rare: the powder-puff flowers of the mimosa family have small pollinia that contain one to three dozen closely packed pollen cells.

The work of replacing atmospheric carbon dioxide with the free oxygen necessary for active animal life was started aeons ago by the earliest green cells, but it remained for the versatile flowering plants to prepare the planet for a latecomer with extraordinarily varied needs: man. Angiosperms supply cereal grains for his breads of many kinds, vegetables for his table, fruits for his health and delectation, spices to stimulate his appetite, fibers for his garments, remedies for his diseases, his finest woods, oils for the paints and varnishes he applies to them, and a thousand other necessities and luxuries to enrich his life. The bounty of the flowering plants is supplemented in small measure by the more ancient conifers, especially as sources of lumber and pulpwood. If all other forms of vegetation were to vanish from the land (except the fungi and bacteria that decompose dead organisms and return their components to the vital circulation), we could continue to live quite well; in the inner tropics of the American continents, we would hardly notice the difference. But, if the angiosperms were to go, we would be lost. Man could not have arisen during the long ages before the advent of the flowering plants. A planet where the vegetation grew no higher than tall shrubs could support a great variety of animals, including the large herbivores, many birds, reptiles, amphibians, insects, and, of course, practically all the marine life that we know; but such a planet could never have produced man.

All animals are indebted to plants for the oxygen they breathe and every scrap of their food, whether they consume plants directly or eat other animals that have nourished themselves on plants. Nearly all terrestrial animals and many that live in shallow water owe to plants the habitat that shelters them, forest, thicket, meadow, or pondweeds. But man, the animal that nature has endowed most generously with mental and physical qualities, has a special indebtedness to plants, especially trees—which, while sheltering and nourishing his remote ancestors, helped prepare them for the evolution of unprecedented mental and manipulative abilities, as well as speech, in a manner that will be explained in chapter 8. If our ancestors had not been frugivorous tree dwellers, we would probably lack the blessing of color vision, which appears to be rare among mammals (see chapter 10). Moreover, plants continue to nourish and protect our bodies, to uplift our spirits with their beauty, and to challenge our minds to understand them.

Of the many questions that we ask about plants, none is more searching, none more difficult to answer, than that about their psychic life. In supporting the whole animal kingdom, plants are magnificently altruistic, but is their altruism quite unconscious? Every gleam of joy, every hour of happiness, that we or any other animate creature has ever known we owe to plants; but do they themselves never feel any satisfaction in their vegetative existence? The protoplasmic strands that penetrate the walls between cells are the only suggestion of a nervous system that plant anatomy reveals; without a more developed system and a central sensorium, an extended vegetable body could hardly have a unitary consciousness, like that of the higher animals. The assertions about their emotional life and sympathetic responsiveness to human moods that we sometimes read are unconvincing, yet plants may not be wholly insentient. If their lives are absolutely devoid of pleasant feelings, why should they live at all, except to support animals, which is hardly an adequate reason? Accordingly, in the absence of positive evidence to the contrary, we may cherish the faith that plants flourishing in the sunshine enjoy a dim or perhaps a vivid sense of well-being, diffused through all their living cells—and this would vastly increase the sum of pleasant feelings in the world, to counterbalance the frequent agonies of the animal kingdom, especially its human fraction.

Under many names, men have worshiped the Sun as a life-giving god. The fertile Earth, often called our mother, has also been deified. Yet, without green plants, neither Sun nor Earth could keep a worm or fly alive. They mediate between the luminary and its satellite and ourselves, giving them all the value they have as sustainers of human life. Plants are our nurses and supporters; and piety, which has been defined as reverent regard for the sources of our being, is more fittingly directed toward them than toward anything else, except the parents who lovingly reared us.

We can understand why certain primitive tribes would not fell a needed tree without apologizing to its indwelling spirit or, perhaps, preparing another abode for it; why the holiest orders of certain religions, such as the Manicheans and the Jains, could neither pluck a fruit nor pick a vegetable nor prepare their own vegetarian meals but depended upon others, less devout, to perform these services for them; why early Buddhist monks and nuns made their pilgrimages in the dry season and remained at their convents during the rainy months, so as not to crush the tender herbage sprouting along the paths. Yet how carelessly do we moderns, whose science has revealed a larger debt to vegetation than earlier generations suspected, destroy Earth's green mantle at an unprecedented rate, to meet exorbitant demands, for gain, and often without any defensible reason! Unfortunately, we cannot feed or clothe or house ourselves without levying a heavy tribute upon the vegetable world, and to permit ourselves to starve would nullify all its gifts to us. But avoidably to cut a tree or pluck a flower reveals lamentable spiritual blindness and ingratitude for inestimable benefits.

To gaze perceptively over a wide green landscape basking in sunshine should dispel the despairing gloom that sometimes overcomes us when we contemplate the countless ills of our turbulent human world or even of the wider animal kingdom, with its ravening predators and subtly debilitating parasites. The source of the landscape's verdure is chlorophyll, the most beneficent matter on Earth, which in millions of leaves is quietly engaged in a wholly constructive process that supports all but a minute fraction of the planet's life—for even in the oceans, where it mostly passes unnoticed, it is active in vast quantities. How can evil predominate on a planet tinted green with chlorophyll, which benefits everything and harms nothing? All human wickedness, all animal savagery, dissipates but a small fraction of the energy that green plants daily supply to the living world. The rest is available for placid and joyous and rewarding activities, and it is within our power to decrease the evil uses while we increase the benign uses of this gift from sunlight and green plants.

5. Animals, Life Groping toward Fulfillment

 A planet covered with lovely and stately vegetable forms, absorbing energy from sunlight, appears to demand something more. Plants are beautiful, but they lack eyes to behold their beauty, to gaze upon the starry heaven above them. They lack ears to hear the liquid murmur of the streams that traverse the forests, the soughing of the wind amid their foliage, or all the melodies that might fill the air. They lack minds to contemplate and try to understand the mystery of their own existence and that of the Universe in which they grow. They cannot move and explore the planet that supports them. With only vegetable life, Earth at its best would be like a garden with no one to enjoy it.

Although any statement we may make about the psychic life of plants is only a conjecture, it seems that, at most, this corresponds to what we feel when we stretch luxuriously in the sunshine, in a state between sleep and waking. Should not the energy that plants store and make available to living cells be applied to more varied activities than plants can perform, to higher modes of awareness than they seem able to achieve? A planet with only vegetable life would realize but a fraction of the values that life can experience. And in a Universe that appears to be striving, however blindly, to bring forth and actualize all the values that it enfolds potentially, should not the high values that plants cannot know be somehow brought to light?

Fully to realize life's potentialities, the best qualities of animals must be added to the plants' basic creativity. Animality must supplement vegetality. To achieve this harmoniously, without destructive conflict, several courses might be followed. One would involve combining vegetable and animal attributes in the same organism. What carnage, what fear and horror and embittered strife Earth would be spared if all animals could synthesize their own food, as plants do! They would have little need to compete except for space, and then only if they became more crowded than at present. Nature, to be sure, made a start in this direction. Some microscopic organisms, such as the euglena, swim through fresh or salt water, propelled by lashing cilia, in the manner of many protozoans, yet they contain chloroplasts and synthesize their own nourishment, so that with equal reason they may be classified in the animal or the vegetable kingdom. Certain simple multicellular organisms, such as the freshwater hydra, are equipped with tentacles that deliver food to a digestive stomach; these organisms, accordingly, are obviously animals, yet green algae that live symbiotically in their cells evidently contribute to their nourishment. A few sponges and sea anemones are likewise green with associated algae.

The newly hatched flatworm *Convoluta roscoffensis* is colorless, but in a few days motile green algae enter its tissues and thenceforth live in a most intimate association with it. At first, the *Convoluta* feeds like other flatworms, but soon its mouth closes permanently and it depends wholly on the algal cells for nourishment. To enable them to carry on photosynthesis, it seeks the sunshine on the beaches of Normandy while the tide is out, then retires into the sand to avoid being washed away by the incoming surf. The worm also loses its excretory organs, probably because its algae use its waste products as a source of nitrogenous food. It becomes so dependent upon the algae that it will die if deprived of them, even if offered abundant food of the kind that it ate while very young. On their part, the algae degenerate until they cannot live outside the worm or reproduce, so that the next generation of worms must be infected by algae of the same kind that have hitherto remained independent. The final mutual dependence of *Convoluta roscoffensis* and its algae justifies the appellation plant-animals. Although this mode of association between photosynthetic cells and organisms that originally lacked them has not been so successful as that of algae and fungi in lichens, it lends credibility to the suggestion that the chloroplasts of plants may be derived from blue-green algae that became established in other organisms.

Nature, then, has experimented with the feasibility of adding the animal's mobility to the plant's capacity to synthesize its own food. Why did it not carry this so promising trend to its ultimate conclusion, creating animals as well endowed with sensory organs, versatile limbs, and receptive minds as the best of them, yet with no need to devour either plants or other animals? We can imagine such an animal equipped with broad green membranes, neatly folded against its body. When hungry, it would spread them leaflike to the sunshine, while it rested or slept or engaged in some agreeable sedentary activity, such as reading or conversing with its fellows. To tide it over stormy periods with little sunshine, the photosynthetic animal might store nutrients in its tissues, as animals of many kinds actually do. Such an animal should be free of most of the toil and stress that have burdened man, who for ages has had to procure food by the sweat of his brow; its life might be very pleasant, with much leisure for intellectual and aesthetic pursuits.

To create an animal, even one as large as man, capable of synthesizing sugar in its own tissues would not, I believe, be impossible for evolution. But organisms need mineral salts as well as carbohydrates, and apparently here is where the difficulty lies. Aquatic, especially marine, animals could absorb all they need through their skins. Terrestrial animals might periodically visit salt licks or the seashore for their minerals. But to only a minority would these sources of salts be within reach; the rest would need to send down roots to draw salts from the soil, as vegetables do, thereby losing the mobility that is one of the animals' most distinctive attributes. Perhaps this is the reason why we do not have plant-animals of advanced organization.

Animals might also have evolved in greater harmony with the vegetable kingdom and each other if plants supplied them with more abundant and nutritious fruits. On the whole, wild fruits (except their seeds) are poor in proteins and not very rich in carbohydrates. Trees and shrubs usually drive hard bargains with the birds and other animals on which they depend to disseminate their seeds, offering them little in return for this essential service. It is surprising to see how eagerly birds sometimes seek watery berries, unpalatable to man and filled with indigestible seeds. Or else the plants surreptitiously attach their prickly or sticky fruits to the bodies of animals, to be carried without recompense. Many, to be sure, dispense altogether with the service of animals and use wind, water, or the explosive dehiscence of their pods to scatter their seeds. If nature had done on a vast scale what man has done on a modest scale, making edible fruits more nutritious and abundant at all seasons, animals would depend more largely upon them, instead of devouring the essential vegetative organs of plants as well as each other.

A beneficent Creator, intent upon bringing forth and making actual all the splendid potentialities of a planet as favored as Earth, while avoiding all contrary developments, would, I believe, have followed one or the other of these alternatives to cover this broad surface with vegetable and animal life in a harmonious union. In the absence of divine supervision, life has had to grope its way forward, often blundering, taking unprofitable directions, making fresh starts, suffering much, yet achieving some admirable results. Multiplying excessively, animals were thrown into relentless competition for food and living space. Some attacked the vegetation, gnawing into the foliage in the manner of many insect larvae, often until scarcely a single leaf remained intact on a great tree. Or they bored into living stems and branches, frequently killing the plant that sustained them. Among the larger animals, the grazers cropped the tender herbage in great quantities, while the browsers stripped shrubs and trees as high as they could reach. Other animals become carnivorous, striking down living victims that they tore and mangled mercilessly. Still others, usually of the smaller sorts, adopted parasitic habits, living on or within the bodies of their hosts, sapping their strength, perhaps causing, in sum, more suffering than did carnivores that killed at a stroke. The privileged planet, which with a beneficent deity to guide evolution might have become the abode of a vast diversity of creatures dwelling in concord, became a place of mixed character, where beauty and ugliness, peace and fear, happiness and horror, mingle together in the most perplexing contrasts.

One kind of excess often leads to another. In the intense competition for food and living space on a crowded planet, large size confers advantages. The big animal needs more of both than the small animal, but it is usually better able to take what it wants. Not only is the large, powerful animal more competent to resist the predators that crave its flesh, but also, in the contest for mates, the big individual can defeat smaller rivals and transmit his bigness to his progeny. The repetition of this process may lead to gigantism, one of the unfortunate trends in evolution. Although its impressive size and strength may win admiration, the huge animal leans too heavily upon the resources of a planet where food and living space rarely suffice for all the creatures that need them. If a terrestrial herbivore, it devours excessive amounts of vegetation and tramples much more. On moist slopes its heavy tread breaks the ground and increases erosion; on dry plains, it impacts the soil. The big carnivore needs an endless supply of victims to satisfy its lust for flesh.

Gigantism reached extremes in the Mesozoic era, when immense carnivorous reptiles such as *Tyrannosaurus*, nearly fifty feet long, dominated the land, titanic herbivores such as the eighty-foot-long, fifty-ton *Brachiosaurus* cropped the vegetation, and a variety of piscivorous giants swam in the seas or glided over them, like the fantastic *Pteranodon*, with a wing span of ten yards. The brains of some of these enormous creatures were narrower than their spinal cords, so that, as a distinguished paleontologist once remarked, they must have had "afterthoughts." About the end of the Cretaceous period, the last of these reptilian monsters vanished from Earth, as though nature had finally decided to terminate a misguided experiment. The mammals that succeeded the huge saurians also showed a tendency toward gigantism, although they never, at least on land, attained such extravagant sizes. By the end of the last glaciation, the mammoths and most of the other gigantic mammals had become or were becoming extinct, leaving the elephants and rhinoceroses as the greatest terrestrial animals and the much larger whales in the oceans.

In his essay on the wilderness, Aldo Leopold wrote: "Only those able to see the pageant of evolution can be expected to value its theater, the wilderness, or its outstanding achievement, the grizzly." To me, this is misplaced admiration. The common habit of applying the epithet "noble" to the biggest, most powerful, and usually most predatory birds and quadrupeds is vulgar and thoughtless. Lions, grizzly bears, and similar animals, especially the males that are the larger sex and, therefore, presumably the most "noble," are commonly selfish and surly—miserable parents who do little or nothing for their offspring. Many a small, weak animal is a far more outstanding or admirable achievement of evolution than these big brutes, by any criterion that we may choose for judging them, except weight and strength.

Compared to a small bird, especially a passerine, the grizzly, the lion, or the eagle is a crude creation. The small bird can do many more things—build a beautiful nest, sing enchantingly, if a migrant find its way over vast distances; it is more pacific, a good parent, and often has a pleasing social life. And it does all this with an exceedingly small fraction of the material and energy that the huge animal needs for survival; it makes much smaller demands upon its environment. The bird is to the grizzly bear as a fine watch is to a stone crusher. And many another animal, among mammals, fishes, and insects, is a more outstanding achievement of evolution than the grizzly. When will man outgrow that primitive mentality—one of the chief causes of his disasters—that adulates power, whether in the form of a callous, egoistic military leader like Alexander of Macedonia or Napoleon Bonaparte or a huge, predatory animal like the lion or the grizzly bear? Perfection is most often found in little things; our planet, the fairest in the solar system, is tiny as celestial bodies go.

As technology advances, it tends increasingly to miniaturize. Among its highest and most recent achievements are miniature models of complicated devices, such as radios and computers, which accomplish most or all that a much larger model of an earlier design could do, with a great saving of materials, space, and energy. Similarly, miniaturization has been an important trend in evolution, parallel to that toward gigantism but much more significant and successful. We find this trend in both the vegetable and the animal kingdoms. Among flowering plants, the small herbaceous species, apparently derived from arboreal ancestors, can do everything a great tree can do except produce wood, which they do not need: they synthesize carbohydrates and proteins, grow, bloom, and mature seeds, just as trees do.

Among invertebrate animals, the social hymenoptera, mostly small to very small even in the insect world, can claim to be considered the most advanced; the ant's minute brain has been called the most wonderful bit of organized matter in the world. Among birds, the mostly small passerines, including the songbirds, reveal important advances over much larger, more primitive types: they build better nests, take better care of their young, sing very much better, and give clearer indications of aesthetic sensibility. Among mammals, man, although far from being the smallest, nevertheless reveals this trend toward miniaturization.

If we compare him with elephants, the larger ungulates, the great carnivores, or the cetaceans, we find that, with a body substantially smaller, he can do a much greater variety of useful and constructive things, and he has greater mental powers.

This beneficent trend toward miniaturization is, unfortunately, accompanied by the tendency to create excessive numbers of the smaller, more economical and efficient models. Multitudes of small animals compete as strongly for the little that each needs as fewer great ones do for the much more that each needs. Crowding and deprivation often prevent the fullest development of their innate capacities. It remains to be seen whether man, to whom the evolutionary tendency toward miniaturization has given a brain large in proportion to his body, has enough intelligence and self-control to avoid the perils of excessive multiplication to which successful animals are exposed.

On a crowded planet whose significance appears to depend, above all, upon the presence of animate creatures that contemplate with gratitude and appreciation everything lovely or sublime that it bears, the most adequate animals appear to be those with the highest ratio of mind or spirit to corporeal mass, so that they may respond to its splendors with love, loyalty, and responsibility, while making the smallest demands upon its bounty. One measure of this adequacy may be the ratio of brain weight to total body weight. But this is not an infallible criterion, for brains differ greatly in quality as well as in size; and too small a brain, or one of poor quality, may still be large in relation to a small body, yet be inadequate to perceive and respond to Earth's beauty and grandeur. Nor is intelligence a guarantee of the desired qualities, for it may be coldly unresponsive to beauty and goodness or even actively malignant. Although we may specify what is most needed to impart a high value to life and the planet that supports it, we do not know enough to point out, without fear of error, the creatures that possess these attributes in fullest measure.

Some kinds of animals have achieved an admirable relationship with vegetation, never injuring the plants that sustain them. Among these are frugivorous birds, such as the lovely tanagers of tropical America, which pay for the fruits that are their mainstay by spreading the enclosed seeds far and wide. Living in perfect harmony with the vegetable world, in pairs or small flocks, they rarely quarrel among themselves; in years of watching them, I have never seen one injure another, hardly ever a brief fight. The large fruit or imperial pigeons of the Orient appear to have achieved a similar relationship with fruiting trees, but, being more elusive, their habits have been less studied. Some of the most magnificent of New Guinea's birds of paradise subsist chiefly upon the fruits of trees and shrubs. The colorful little manakins of the American tropics are also largely frugivorous and peaceable. Likewise hummingbirds, whose diet consists of the nectar of flowers supplemented by tiny insects that they catch in the air or glean from leaf and stem, pay for their sweet drink by pollinating the flowers— and they are incapable of harming the plants that nourish them. The almost wholly vegetarian parrots are in a different category; preferring the more sustaining embryos to fruit pulp, they destroy countless seeds and probably take little part in their dissemination. Exceptional in this family are the lovely brush-tongued lories of Australia, which drink nectar and are true pollinators. Among insects, adult butterflies are as innocuous to vegetation as to man, although in the larval stage they can be destructive to foliage. Some species of nectar-sucking bees hardly harm plants at any stage of their lives.

Is it only an accident that these creatures which have achieved the most harmonious relations with vegetation, as with the rest of the living world, include such a high proportion of the animal kingdom's most beautiful members? Tanagers, birds of paradise, fruit pigeons, manakins, hummingbirds, the mainly frugivorous trogons, brush-tongued lories, butterflies—what a brilliant galaxy! Must not the harmony of their existence be somehow related to their visual beauty? Compare their splendor with the dull or dingy coloration of the most highly predatory birds and mammals, the hawks, owls, and terrestrial carnivores.

As we survey the tremendous variety of animals on land, in the air, and in the water, from scarcely visible mites and insects to whales thirty yards long and birds whose wings spread twelve feet from tip to tip, some insistent questions cross our minds. What have they all contributed to their planet that plants themselves could not give? What higher values have they brought to it? A minority have become serviceable to plants as pollinators and carriers of seeds, but many are harmful to vegetation. The animal kingdom has added a vast amount of motion to Earth, but of that there was no lack before life arose upon it, in the revolving planet itself, in wind and waves, in rivers and ocean currents, in the invisible darting atoms and molecules. What matters is not activity itself but what activity accomplishes.

Many animals have substituted for the plants' steady, quiet creativity brief intervals of intense predatory ferocity, between which they rest in almost vegetative somnolence. These hardly raise creation to a higher level or make their planet more attractive. More pleasing are the animals that show some constructive ability, building nests that are often beautiful, curious, or surprisingly complex, such as those of many birds and social insects and not a few solitary species. Such fabrications, including the most intricate that men have made, hardly approach, in complexity and close integration, the bodies of the animals that construct them; but they appear to require a certain deliberate dedication, or at least they provide an opportunity for such dedication; and this addition to the basic creativity of organic growth, photosynthesis, and similar biologic processes has great potentialities for our planet.

Plants cooperate passively or, one might say, unintentionally, as when the many diverse trees and other growths of a tropical forest collectively prepare and maintain the conditions indispensable for the perpetuation of this same forest from century to century. Animals cooperate more actively, and perhaps more intentionally, than is possible for plants, in finding food, building nests, rearing young, and protecting each other from danger. In the social insects, especially those more advanced species in which a multitude of biologically incomplete individuals, the workers and soldiers, rear and guard the progeny of one or several fertile females, cooperation appears to have become compulsory, like that between the cells of a living body—unfortunately, we cannot penetrate the psychic lives of creatures so different from ourselves.

Among the warm-blooded vertebrates cooperation seems, at least to our inevitably biased vision, more deliberate and purposeful. Thus, when we watch fledgling, yearling, or even older birds, in many cases still sexually immature, assisting their parents or perhaps a neighbor in feeding and protecting a younger brood, we suspect some conscious motivation for such helpfulness and psychic bonds between the members of the cooperating group. Birds who have lost their own young sometimes assist more successful parents, of their own or other species. Monkeys, apes, and other mammals frequently take a helpful interest in their neighbor's dependent young and may try to raise them if they are orphaned. All such cooperativeness and concern for other individuals mark a momentous advance in the living world.

More than any other animal activity, care of dependent young, such as occurs in certain insects, fishes, and frogs, in all mammals, and, above all, in hardworking avian parents, promotes affective bonds. This was evidently the context in which love, properly so-called, was born. When, as in many avian species of mild climates, the parents remain in inseparable pairs throughout the long season when they are sexually inactive, the kind of attachment that binds the parents to their young appears to have been transferred to the partners who helped each other cherish those young. Here, in any case, is the kind of situation in which love that is more than a strong transient passion is likely to arise and grow. Once its seeds have been sown, love tends to be expansive, spreading ever farther from its source. And one of the greatest contributions of animals (including man) to their world is love, of family, of social companions, of all things good and beautiful, of the exceptional planet that supports them, of the ideal perfection called God.

A planet with magnificent scenery, covered with lovely vegetable forms beneath a sky delightfully azure and white on fair days and jeweled with countless stars at night, would be incomplete without beings to respond with grateful appreciation to all this sublimity and beauty, to wonder at and try to understand their world and themselves. This is the most compelling reason for believing that it needs animals with eyes and ears and minds to supplement the beneficent constructive work of plants and make full use of the energy they provide for vital activities. What do we know about the aesthetic sensibility and intellectual curiosity of non-human animals?

We can dismiss the latter in a sentence. Although many animals, especially birds and mammals, evince curiosity, it seems always to be directed toward what is immediately present, is always superficial, and is never, as far as we can tell, concerned with the deeper causes of phenomena; it never merits the designation "scientific" or "philosophic." With aesthetic sensibility, the case is different. Writers have imagined that in the future people will be entertained by a "symphony" of scents. We cannot dismiss the possibility that mammals, some of which have noses incomparably keener than ours, delight in combinations of scents no less than we enjoy compositions of sounds and colorful forms; but, since we cannot accompany them into their olfactory world, to pursue this matter further would be profitless. This leaves birds, who in certain ways resemble us more closely than any other animals, not excluding our brother mammals, as the creatures most likely to have an aesthetic sense.

It used to be thought that birds sing because they are happy, for, as Plato remarked long ago, no bird sings when it is wet and cold. When it was recognized that birds sing to proclaim possession of a territory and to attract mates, there was a tendency to deny that they sing for enjoyment. We who embellish our useful wares should know better than to assert that the utilitarian and the aesthetic are mutually exclusive. Some birds sing, alone or in flocks, when neither defending territory nor trying to win mates. Some imitate not only their songful neighbors of other species but other natural or manmade sounds, which is strong evidence that they are interested in sound for its own sake. They display a certain virtuosity in perfecting their own songs and their imitations, leading one writer to declare that in imitative singing birds give their clearest indication of intelligence. Since a distinctive squawk might also serve to advertise territory and need of mates, why should some birds sing so enchantingly if they and their partners care nothing for melodious sounds?

For the bright plumage of many birds, often gorgeous in total effect and intricate in detail, the most convincing explanation yet advanced is Darwin's theory of sexual selection. This holds that these birds, usually males, have become so handsome as a result of preferential mating, the females responding to and choosing as their transitory or permanent partners the individuals who most impress them by the elegance of their attire or the vigor of their display; and this in turn implies that birds have something difficult to distinguish from an aesthetic sense.

Perhaps the most convincing evidence of birds' aesthetic sensibility is provided by the bower birds of Australia and New Guinea. These relatives of the birds of paradise build, on the ground, short avenues of interlaced twigs, pillars of the same material heaped around saplings, or conical, wigwamlike huts of slender orchid stems. Some species paint the interior of their bowers with charcoal or vegetable juices. They decorate their bowers with shells, pebbles, or a miscellaneous assortment of small, colorful articles. The Satin Bower Bird prefers ornaments that are blue (the color of the adult male's plumage) or yellow-green (the color of the females and immature males). The most wonderful of these constructions is that of the Vogelkop Gardener Bower Bird, the most plainly attired of them all, who evidently substitutes a colorful garden for personal adornment as a means of attracting females. On a mossy "lawn" that he lays in front of his wigwam, he arranges little heaps of fruits, flowers, fungi, and other small objects, usually each pile of a single color, which may be red, orange, yellow, blue, or violet. When the flowers wither or the fruits decay, he removes them and brings others. Like most of the bower birds, he is an accomplished mimic.

From this rapid survey, we may conclude that animals add, to the accomplishments with which plants had already endowed Earth, some ability to construct things other than their own bodies; active cooperation among pairs or larger groups and, above all in certain insects, a complex social structure; parental care, which gave rise to love; a capacity for strong personal attachments, especially among birds; and at least the first glimmerings of aesthetic appreciation, again most evident in birds. Indeed, all these promising developments appear among birds more clearly than among any other major group of animals, except that social integration is carried farthest by insects.

Nevertheless, when we view the whole animal kingdom, excluding man, in its vast extent, these developments are of sporadic occurrence, and rarely have they been carried far. They are only hints at possibilities that await realization. In their totality, they fail to fill the need or the opportunity for grateful appreciation of all Earth's beauty and grandeur, for wonder and some understanding of the cosmos. Indeed, the animal kingdom vastly increases this need, for it adds to all the beauty of the vegetable kingdom an immense variety of graceful forms, lovely colors, and fascinating ways of life. Animals add immeasurably to the interest and attractiveness of our planet. Yet their advent upon Earth was no unmixed blessing. While many animals attract, others repel us by their predatory ferocity, even when, with our mechanical advantages, we have no reason to fear that they will harm us. Earth has no more distressing spectacle than that of a predator suddenly striking down some defenseless creature innocently singing or attending its young, no sight more pitifully repulsive than the hideously mangled remains of what, a few hours before, was a beautiful animal enjoying its life.

This leaves man as the only animal which seems capable of contemplating with absorbed interest the superb spectacle of a planet covered with life, of somewhat adequately appreciating all its beauty and grandeur, of trying to understand it, and of gratefully acknowledging his privilege of living upon it, thereby increasing immeasurably the significance of his own life and that of the world which bears him. Yet how few of us appreciate our planet as we should, how few are thankful for the privilege of dwelling upon it, how few feel responsible for preserving its fruitfulness and beauty, how few try earnestly to understand! And, even for the best of us, how rare and brief are the intervals when we are fully alive to the splendor around us, when we think clearly about the meaning of our presence here!

All this emphasizes, once more, a general feature of the world process: the tremendous difficulties, the immense travail, that each major advance has involved. So many stars, so that a few might illuminate planetary systems! So many planets, so that a few might bear abundant life! So many animal species, so that a few might realize the highest potentialities of animate life! So many people, so that a few might become somewhat adequately grateful for their privileged situation on a privileged planet and awake to the responsibilities that correspond to their unique status!

6. Insects, the Enigmatic Throng

 As animals, insects might have been considered in the preceding chapter, but for several reasons they deserve a chapter to themselves. Of the estimated one million named species of animals on land and sea, about three-quarters are insects, and many new species are described each year. Since scarcely any of these insects live in the oceans, they account for an even larger share of the terrestrial faunas, perhaps as much as 90 percent of the total number of species and several times as many individuals as all the other animals together. In size, modern insects range from scarcely visible beetles and parasitic wasps to the Brazilian moth *Erebus agrippina*, with a wingspread of eleven inches; but some dragonflies of the Carboniferous period had nearly thrice this span.

No matter where we live, insects are the creatures that most often come into contact with us, accidentally or to suck our blood. They are regarded as man's chief competitors for dominance of the planet; if our criterion of biological success is number of individuals, they are far ahead of us. By their attacks on crops, buildings, and household goods of many kinds, they cause more economic problems than all the other animals together. For this reason, no less than for their intrinsic interest, insects have probably received more study than any other division of the animal kingdom, and an immense literature has grown up about them. Nevertheless, when we broadly survey the life of our planet, the significance of each of its major branches and their mutual interactions, insects remain a major problem, an outstanding enigma. In the preceding chapter, we considered how

plants and animals complement each other by their different capacities and together give the planet a completeness
and a significance that it would lack if it bore plants alone.
There we were concerned chiefly with the vertebrates. But
how do the teeming insects fit into the picture?

It is not difficult to account for the abundance of insects,
as species and individuals. The brief life cycles of many
kinds have permitted an enormous number of generations
to succeed each other, in many lineages, since the first
appearance of insects in the geological record—in the Mississippian or early Carboniferous period, about 340 million years ago. A short life cycle favors rapid evolutionary
change and, incidentally, its study by man; the fruit fly,
Drosophila melanogaster, has contributed more to our understanding of the genetic basis of evolution than has any
other organism. The capacity for rapid evolution has enabled insect stocks to diversify along with the plants on
which they subsist. Since many insects specialize upon a
certain group of closely related plants, upon one particular
species, or upon a particular section of a certain plant, such
as its leaves or its seeds, and since, moreover, many of these
vegetarians have their own special insect parasites, the
huge number of vegetable species on our planet is largely
responsible for the much vaster number of insect species.
Indeed, the coleoptera or beetles, many of which feed upon
plants, account for somewhat over a third of the total number of species of insects.

Although many insects live in or on fresh water at some
stage of their development or even throughout their life
cycle, and a number of species favor highly saline lakes or
pools, very few inhabit the oceans. Water striders of the
genus *Halobates* are truly pelagic, skipping over the surface
of the open seas far from land, breathing air, and eating
floating dead animals. A number of chironomid flies live
completely submerged in sea water; the larvae and wing-
less, legless females hide in tubes that they build between
the leaves of underwater plants, and only the males swim

actively. When one recalls that the arthropods, the great zoologic phylum to which the insects belong, includes the crustaceans, so well represented in the seas by shrimps, crabs, lobsters, and their relatives, the failure of more than a few of the so adaptable insects to take to the seas is the more surprising and has not been satisfactorily explained. Possibly the chief obstacle to insects' colonization of the oceans is the difficulty that such small creatures must experience in breathing air in turbulent waters.

Although certain insects make themselves so obnoxious to ourselves and to other animals by sucking our blood, only a small minority misbehave in this fashion. Many more prey upon or parasitize other invertebrates, chiefly insects. About half of all the species of insects feed upon plants, including fungi, or their products, and this is their chief significance in the economy of the living world. They mediate between vegetation and other branches of the animal kingdom. Countless insects die each day to feed spiders, which in turn may nourish other insects, notably wasps, many of which provision their brood cells with paralyzed spiders that are slowly devoured by the larval wasps. Apparently, however, the net advantage in the strife between the hexapod and the octopod worlds is on the side of the spiders, whose insidious webs ensnare insects in greater numbers than spiders are captured by insects.

The chief importance of insects in the food chains involves supporting the life of vertebrates. Trout and other freshwater fishes, frogs and toads, small lizards of many kinds, and the multitudinous nocturnal bats all depend largely upon insects for food. Anteaters, pangolins, shrews, moles, certain lemurs, and other flightless mammals feed upon insects, and they supplement the diets of largely frugivorous monkeys and apes. Even man, especially in his less fastidious primitive state, varies his meals with insects. But it is above all the birds that catch insects and help balance their prodigious power of reproduction. Swifts and swallows capture them while circling tirelessly in the air, and they eat scarcely anything else. Flycatchers of many kinds seize them chiefly on short sallies from lookout perches. Wood-warblers and vireos glean them from foliage. Woodpeckers extract them from infested trunks and branches. Antbirds and others catch those that try to escape the hordes of army ants that flow over and through the ground litter of tropical forests. Gallinaceous birds uncover them by scratching the ground. Even mainly frugivorous birds, such as the tanagers, vary their diet with insects. These and other generally vegetarian birds feed their young with a larger proportion of insects than they themselves eat, to supply proteins for their growing bodies. Only a few birds, including certain cotingas, seed-eating finches, and the strange Oilbirds of South American caves, raise their nestlings on a wholly vegetarian regimen. Insectivorous birds commonly include in their diet spiders and other small invertebrates that have devoured many insects. Other birds eat lizards and frogs whose flesh was nourished by insects. Directly or indirectly, insects support a substantial portion of the avian world that so embellishes our planet.

Insects concentrate the foods that plants manufacture, especially proteins, making them more readily available to birds and other animals. In doing so, they may seriously injure vegetation, devouring leaves, gnawing into stems and roots, ruining seeds. A botanical collector in tropical forests is often exasperated by the difficulty of finding a few twigs, even from a large tree, with leaves that have not been disfigured by gnawing insects. In such forests—composed of a bewildering variety of trees—insect damage, although widespread, is rarely severe enough to affect the aspect of the woodland as a whole. It is otherwise in temperate-zone forests where few kinds of trees grow, especially in coniferous forests, which may be devastated by periodic attacks of such pests as the spruce budworm.

Just as a hungry world is increasingly recognizing that, as converters of vegetable products for human consumption, cattle and other domestic animals are extremely inefficient, so it appears that, as concentrators of vegetable proteins for birds and other small creatures, insects are inefficient. A biological engineer who wished to cover his planet with flourishing vegetation that supported, with a minimum of waste, a great variety of colorful, songful birds would omit plant-devouring insects and design his greenery to yield a greater profusion of sustaining fruits. If plants could devote to the formation of such fruits the nourishment they now lose to insects, they and the avian world would benefit.

More beneficial to vegetation are the nectar-sucking and pollen-eating insects that fertilize flowers (some of which, such as butterflies, are leaf gnawers in their larval stage). Although wind-pollinated flowers, including those of many grasses and temperate-zone trees, are independent of insects, the brighter blossoms owe their existence to these winged creatures. Colorful, fragrant flowers and their insect pollinators evolved together, the plants producing ever more colorful blossoms, ever more enticing scents, as they competed for the services of insects that were becoming more discriminating than the primitive pollinators, such as the pollen-eating beetles. Since the vision of insects extends into the ultraviolet region of the spectrum, invisible to us, but may be insensitive to much of the red, as in bees, they often see flowers otherwise than we do; a photograph of a flower taken with ultraviolet light may reveal a pattern that our unaided eyes fail to distinguish. But most pollinating insects respond positively to the same floral scents that we find delightful. Only a small minority of flowers, usually brownish or livid in color, including the fleshy African stapelias and certain orchids and aroids, emit a putrid odor to attract carrion-feeding flies as pollinators.

In the tropics, and to a much smaller degree in the temperate zones, birds are important pollinators—hummingbirds and honeycreepers in the New World; sunbirds, honey eaters, and lories in the Old. Flowers that depend upon avian pollinators are often brilliant but frequently devoid of fragrance, as birds are rather insensitive to scents. They are often tubular or trumpet-shaped, without the wide-spreading petals that serve insects as landing platforms; or else they are clusters of exserted stamens, the so-called bottlebrush type of flower. In the New World, many of the flowers specialized for pollination by hummingbirds have been derived from insect-pollinated ancestors. Probably, if the older insects had not begun the work of pollination, the birds would never have found in flowers the rich nectar that enticed them to participate in this necessary activity. We must thank the insects for the colorful flowers that adorn our planet and for the delightful fragrances that many of them pour upon the air. And, if there would not be so many kinds of insects if there were not so many species of plants, it is equally true that there would not be so many kinds of plants if there were not so many species of insects. The latter, by their pollinating activity, have greatly increased the diversity of floral structures upon which botanists so largely depend for classifying plants. The largest plant families, the composites, the orchids, and the peas (Leguminosae), are mainly insect-pollinated.

Not only indirectly, through their interactions with plants, do insects embellish Earth. Many are themselves beautiful. One thinks first of the wide-winged butterflies, displaying elegant color patterns as they float lazily in the sunshine, sip the nectar of flowers with their coiled proboscises, or settle in flocks upon moist sand. Next in beauty come the beetles, with shards of many colors, golden, shining green, burnished copper, red, yellow, or blue. Some of the bees, especially the euglossids, are intensely metallic green or blue. Those who peer intently into the herbage may delight in less conspicuous hexapods, especially the little membracids or treehoppers, often so curiously shaped, so boldly patterned in bright hues—and so quick to leap out of sight.

A few among the hosts of insects are skillful builders. The most imposing examples of insect architecture are the work of termites, especially those of arid regions. On the

West African savannas they erect termitaries that may become twenty feet high by a hundred feet in diameter. Some of the smaller termites' nests are more elegantly symmetrical, with elaborate arrangements for ventilation and temperature regulation. In all parts of the tropics, termitaries are appropriated by nesting birds, including species of parrots, trogons, kingfishers, puffbirds, jacamars, and woodpeckers, who excavate chambers in the midst of the hard material and there lay eggs and raise broods, while the termites close off the galleries that abut upon the birds' quarters.

In the New World tropics, the outstanding examples of insect architecture are constructed by social wasps. Composed of a paperlike material, attached to tree trunks, walls, rocks, or the undersides of leaves, these vespiaries are of the most diverse sizes and shapes and are often of delicate beauty. In the walls of some nests are innumerable tiny windows, each closed by a thin transparent membrane, to illuminate the interior. A few of the social wasps build shapely urns of clay, treated to resist the hardest downpours. Within all these nests are sheets or tiers of the usual symmetrical hexagonal cells. Since, unlike some of their hot-tempered northern relatives, these tropical wasps rarely sting unless they are rudely disturbed, their curious constructions may often be examined closely without danger. Often a bird builds a covered nest close to one of the larger arboreal vespiaries, whose inhabitants may protect the eggs and nestlings by driving away predators that shake the supporting limbs.

These termites and wasps, along with the ants and bees, reveal the extremes to which social integration may be carried, even in a class of animals, such as the insects, in which organized societies of any kind are exceptional. In some of the hymenopterous societies, as among the termites, we witness the strange phenomenon of fertile females unable to attend the young they produce in vast numbers, males who do nothing but inseminate the females, workers who nurse young they cannot beget, soldiers so specialized that they can do little but defend the colony. Functions that among ourselves, as among birds, can be adequately performed by a single male and female cooperating together are, in the more elaborate insect societies, divided among four or five structurally different castes. No individual is complete and self-sufficient, but together they compose a sort of superorganism, so efficient that, as in the ants so extremely abundant in warm lands, it may justly be called the most successful form of terrestrial life. Not only do these amazing social insects provide homilies for moralizers, they furnish an example—admirable or horrible, according to one's taste—of what humanity may become, if those who propose breeding special types for particular occupations ever succeed in controlling our destiny.

Insects share with man the distinction of being the only agriculturists in the animal kingdom, and they long preceded us in the praiseworthy endeavor to produce one's own food instead of relying upon the bounty of unassisted nature. The hexapod agriculturists include certain ants widespread in the warmer parts of the New World, some termites of the Old World tropics, and the widespread ambrosia beetles of the families Scolytidae and Platypodidae. These beetles live in tunnels that they bore into the wood of freshly dead or dying trees, where they grow the fungus that is the only food of their developing larvae. Each species of ambrosia beetle lives symbiotically with one particular kind of fungus, of which the emerging adults carry spores, in a bunch of hairs on the front of the head or in the stomach, to the new burrows that they will soon excavate, thereby maintaining the culture from generation to generation. The larvae inhabit pockets or chambers along the burrows and are fed by the adults in a manner reminiscent of the social hymenoptera, although special castes are not known to occur among these beetles.

All the unrelated insect farmers are monoculturists that specialize in growing the fungus on which each species subsists. Best known are the leaf-cutting or parasol ants of the genus *Atta*, whose long columns of small brown workers, each carrying above its back a flat piece of green leaf that it has just cut from some tree or lower plant, are a familiar sight in tropical America. If you follow the column moving along the narrow, bare path that the ants have made, you may watch it pour into an opening in a mound, which may be of any size up to about seven or eight yards in diameter. In chambers beneath the mound, the leaves are cut up and heaped in masses that remind one of coarse, decaying sawdust. This is the medium on which a pure culture of a special kind of fungus is grown, carefully attended by workers of the smallest size. The fungal filaments that penetrate the leafy mass in all directions bear tiny stalked knobs, called bromatia or, more familiarly, kohlrabi bodies, which form the ants' diet. On her nuptial flight, each female parasol ant carries, in a special pouch in her throat, a mass of fungal threads which, if she is successful in starting a new colony, will seed its fungus garden. Unfortunately, to grow their food these ants need great quantities of fresh green leaves, and they too often cut them from cultivated plants, sometimes quite defoliating them and thereby coming into conflict with human gardeners and agriculturists, who might otherwise befriend them as fellow toilers in an honorable occupation.

Among the most admirable products of evolution are the bees, especially the small, stingless bees of the meloponine tribe, which live in populous colonies in trees or in the ground and contribute greatly to the exuberant plant life of the tropics by serving as its indefatigable pollinators, but pollinators almost incapable of injuring it. Students of the hymenoptera appear to agree that solitary, predatory wasps, in particular the branch of the group that gave rise to the Specidae, were the ancestors of bees. The fact that predatory insects who contribute nothing to the vegetable world could evolve into harmless creatures who have become indispensable to it should give hope to mankind, which has also been passing through a predatory stage but for thousands of years has been striving, through its spiritually advanced members, to rise above it and achieve more harmonious relations with the rest of the living world, including the vegetation that supports us. What insects could do, humanity should be able to do—and in far less time than the millions of years that the hymenoptera probably needed to effect the transformation, for *Homo* appears capable of evolving more rapidly than other animals.

The purpose of this chapter is not to write the natural history of insects, which even in summary would require a thick volume, but to remind us of some of the salient facts that should be borne in mind when we try to assess their role in the living world. Would our planet be a better place for other forms of life if this most prolific class of animals had never evolved upon it? Would our own lives be enriched or impoverished by their total extinction? We might review the insect families one by one and decide that, if we could, we would retain this and exterminate that; but, since they are all here together, the parasitic dipterans, the industrious bees, the elegant butterflies, the good and the bad, let us judge them as a whole.

How would the vegetable kingdom be affected by the disappearance of the insects? Many a tree, shrub, or herb that must now carry on photosynthesis with leaves perforated and mutilated by insects, or that has been quite stripped of its foliage, might have its full complement of perfect leaves, enabling it to manufacture more food and grow more quickly. No insect galls would form excrescences on leaf or stem. A higher proportion of seeds would be sound and germinate. On the other hand, many plants that depend largely or wholly on insect pollinators would set fewer seeds or none at all. Earth's vegetation might be poorer in variety, but not for that reason unable to cover the land with greenery as dense as climate and man permit to flourish.

With the disappearance of the insects, whole families of birds would vanish from Earth, unless some substitute food could be promptly found for them. Bats might be even more severely afflicted, but their loss would not be so widely lamented by man; especially in the tropics, they can be troublesome animals, fouling buildings where they hang up by day or between nocturnal excursions and, unlike birds, doing nothing to brighten our world with color and song. Many flightless insectivorous mammals, along with lizards, frogs, and toads, would go hungry without insects. Even freshwater fishes might feel the pinch. It is certain that the removal of the insects, which are rarely a conspicuous feature of the landscape, would cause great changes in many of Earth's environments.

And how would we ourselves be affected? We would be relieved of vermin that persists in infesting our homes, despite our best efforts to keep them clean, of the termites and ants that riddle their timbers, of the flies and mosquitoes that suck our blood, of the major part of the pests that attack our forests and cultivated fields and cause us to pollute our planet with dangerous pesticides, of the chief despoilers of stored grains and other foods, and of the vectors of a number of debilitating or fatal diseases. But we would miss the butterflies that brighten our sunny meadows, the chirping and humming of the innumerable small creatures that form a companionable undertone to our walks through the fields, the fireflies that on warm summer nights dot the landscape with points of light like the stars above. We would regret the loss of the lovely flowering plants that might disappear without their insect pollinators, while orchardists and farmers would find the yield of various crops reduced if deprived of their little winged cooperators. Naturalists would deplore the loss of things to study, for, however troublesome they may be, insects always fascinate the inquiring mind. Finally, we may raise, without trying to answer, the question whether, without pollinating insects, warm forests would have become rich and varied enough to support our distant arboreal ancestors, who, during a long age in them, laid the indispensable foundation for man's evolution. Although undoubtedly we would gain in many ways if our great competitors, the insects, were completely removed from this planet, perhaps we would lose more.

But before deciding to exterminate the insects, if this were in our power, thoughtful people might ask some searching questions about them. Do they increase the significance of the planet by enjoying their existence upon it? Is their psychic life more intense or more varied than that of plants, or should we regard them simply as more mobile examples of vegetative existence, serving to pollinate flowers and concentrate nutrients for animals that are more richly endowed, but in themselves hardly lifting sentience to higher levels? Do bees and butterflies enjoy the floral colors and scents that delight us, or do they react to them without feeling? Have ants and wasps any affection for the companions with whom they exchange food or for the homes that they so valiantly defend? We know that the senses of insects are marvelously keen, enabling them to detect the most minute traces of substances, and that their instincts or innate patterns of behavior are often wonderfully complex. But all this tells us nothing about their psychic life; man can now make mindless machines that are equally sensitive or accomplish comparably intricate feats.

From the way that insects sometimes continue their activities immediately after a severe mutilation, we suspect that they are rather insensitive to pain, but this certainly does not preclude their feeling pleasure; it may be that they are luckier than we are, being so organized that their nervous system registers only the more pleasant side of sensation. Without knowing the quality of their inner lives, without knowing what it is like to be an insect, we do not know the most important thing about them. And, as so often happens when we ask the most searching questions about ourselves and our world, the best answer that we receive is no better than a surmise.

One of the most revealing facts about the psychic life of social insects involves the relation of the workers to the queen mother and the larvae that they support. We are apt to think of the worker class among bees, wasps, ants, and termites as patient, self-sacrificing drudges, who wear out their strenuous lives rearing countless young that they did not beget, without the slightest reward for their devoted toil. But this is not quite accurate. The larvae of a variety of social insects have, in or around their mouths, glands whose secretion is highly attractive to their nurses. After feeding their dependents, the workers lick up this substance with evident satisfaction. It is as though a farmer sold a bushel of potatoes to buy a few ounces of candy: the larva receives more nourishment than it gives; the attendant apparently gains in pleasure. Similarly, the workers of certain species of termites so relish the secretion of their queen, whom they must feed liberally to insure the production of an endless stream of eggs, that they sometimes tear pieces from her epidermis, leaving small scars. Such exchange of food or gustatory satisfactions among insects, serving as a social bond, was designated "trophallaxis" by the great entomologist William Morton Wheeler, who felt confident that these insects "give unmistakable evidences of memory, appetites, emotion, imitation and a feeble intelligence."[1] Apparently, the toilers in these insect societies, like workers everywhere, demand their little pleasures.

The insects are many times more numerous in species than the vertebrates, and the members of this great class differ among themselves no less than man from snakes and fishes. Their psychic life, could we fathom it, would no doubt prove to be no less diverse than their structures and habits. Our dense ignorance of this aspect of the most abundant of terrestrial animals reminds us, once again, of the severe limitations of our knowledge.

No less puzzling than the inner life of insects is man's attitude toward them. The laws of Asoka, who reigned in

[1] William Morton Wheeler, *The Social Insects: Their Origin and Evolution* (London: Kegan Paul, Trench, Trubner and Co., 1928), p. 225.

India from about 272 to 232 B.C., afforded some protection to insects, as well as to a surprising variety of other creatures. The devout Jain householder cooks and eats his supper before nightfall, so that he may see and remove any tiny creatures that fall into his food; the Jain monk is careful not to crush them when he walks or sits. Taoist texts of ancient China admonish the reader to respect the teeming insect life. But, in the modern West, insects are neglected both by protective legislation and by the propaganda of the humane societies; scarcely any voice is raised in their defense. As far as I know, only butterflies have societies for their protection and reservations to preserve their rarer species, simply because people enjoy their beauty.

Doubtless, to care about insects requires wider sympathies, and a more active imagination, than to care about dogs and cats and horses or even about wolves and lions. By broadcasting powerful toxins, we destroy many kinds of harmless or economically beneficial insects along with the injurious species against which the poison is aimed. By exposing unnecessary lights at night, we lure many innocuous insects to their doom—in a thousand ways, we reveal our callousness toward these small creatures. How can we pretend to have reverence for life when we have none for the most numerous class of living things on our planet? Even when compelled to destroy some of them to safeguard our vital interests, anyone with a proper respect for life does so with reluctance and regret.

7. Evolution, Life's Gambling Advance

 In the primal seas, ages ago, life arose when dispersed molecules came together in a complex pattern able to reproduce itself. The earliest living things were certainly very simple, of microscopic size or perhaps, like filterable viruses, too small to be detected with an ordinary microscope. They were doubtless all much alike, for, even if life arose repeatedly, the organisms that survived to leave descendants in the present age were basically similar, as seems evident from the fundamental biochemical uniformity of all contemporary animals and plants.

Today, some 3,000 million years later, Earth bears about a million species of animals and between a third of a million and half a million species of plants. In size, these organisms range from submicroscopic viruses (if these are admitted to be alive) to redwood trees nearly four hundred feet high and whales ninety feet long. Their diversity in form is indicated by the number of species, as only exceptionally are two species so similar in appearance that they can be readily distinguished only by physiological or behavioral differences. They swim, they creep, they walk, they fly, they burrow underground, or they stand immovably rooted in the soil. They cover the land from the seashores to the edges of the permanent snowcaps on the highest peaks and even beyond, from the warm tropics to polar ice fields, from dripping rain forests to arid deserts; they swarm in the oceans from the sunny, wave-tossed surface to the dark, cold, abyssal depths, where they survive under tremendous pressure. How did creatures so large and complex arise from such small and relatively simple ancestors? How did primitive uniformity produce such vast diversity?

This diversification becomes even more puzzling when we recall that the very first requisite of living organisms was the capacity to produce accurate copies of themselves. Until the complex molecules, the forerunners of life that arose in the primal waters, could replicate themselves, their patterns existed precariously, in constant danger of extinction. When they succeeded in multiplying copies of themselves, life may be said to have become firmly established, for if some examples were destroyed others might persist. Today, the perpetuation of any animal or vegetable species is strictly dependent upon this capacity to reproduce the existing type.

DNA and Mutation

The form and functioning of an organism are controlled by the DNA molecules in its nuclei, by means of the enzymes that they synthesize. The long double helix of deoxyribonucleic acid, sometimes called the coil of life, is able to replicate itself with great accuracy, so that, when a cell divides, the daughter cells nearly always receive exact copies. As the organism reproduces, the pattern of the DNA is transmitted from generation to generation. The stability of the molecules that determine heredity is attested by the fact that many types of plants and animals have persisted with little change for millions of years.

The replicating mechanism of the DNA helix is so precise that, if nothing interfered, it might continue indefinitely to multiply copies that resemble each other as closely as pages printed from the same stereotype, in which case species would be as fixed and stable as Darwin's early opponents wished them to be. However, these complex molecules are not immune to accidents. Hard radiations, including X rays and gamma rays, penetrating the protoplasm in which they are embedded may knock about the atoms, changing the structures of the helix at one or more points. High temperatures, which increase the mobility of atoms, may also cause alterations in their arrangement.

As we would expect, each alteration of the controlling pattern causes some change in the structure or functioning of the organism. These changes, known as mutations, are of endless diversity; although some features of some organisms are more stable than others, no detail of any plant or animal appears to escape from them indefinitely. Size, rate of growth, relative proportions of organs, color, abundance and form of hairs or scales, chemical composition of tissues, metabolism, longevity—all these characters and many more are subject to mutation.

When mutations are artificially induced in the laboratory or by the radioactive fallout from atomic bombs, most turn out to be harmful, and the same appears to be true of those occurring naturally. Indeed, since they result from the accidental disruption of a coherent pattern, it would be surprising if many of them proved to be beneficial; they might be compared to the careless mistakes of a typesetter, which scarcely ever improve a well-constructed sentence. Yet, upon these obscure accidents at the molecular level, all evolutionary advances depend. If the DNA molecules were permitted to replicate themselves without interference from age to age, the living world would be far simpler and less diverse than we find it today. Perhaps life would never have emerged from the water, and we would not be here to enjoy and study it. Just as we, their descendants, often need some accident or disruption to jolt us out of a comfortable existence and make us strive to improve our situation, so might primitive organisms have remained always in somnolent tranquility if they had not been occasionally stung by hard radiations or jarred by high temperatures.

Species that differ as much as, let us say, man and horse nevertheless have so many features in common that they are evidently descended from a very remote common ancestor, a primitive mammalian vertebrate. The wide divergence between man and horse results from the accumulation, in each stock, of a huge number of mutations, carrying the two lineages in different directions but making of each well-integrated, viable animals. Since mutations occur at random, and most are useless or even harmful, it is most unlikely that all those necessary to make a man, or a horse, out of a small ancestral mammal would occur in a single line of descent and at such times as to complement each other. For evolution to proceed, some arrangement to bring favorable mutations together in a coherent pattern and to eliminate unfavorable mutations was indispensable. This is the function of sexual reproduction.

Genes, Chromosomes, and Sex

Sex, itself a product of evolution, is one of its most surprising, most improbable developments. Both plants and animals have many other methods of reproducing—by simple fission, as in many unicellular organisms; by budding, as in the hydra; by spores and gemmae, as in many of the simpler plants; by larger detached parts of the parent, such as the bulbils, tubers, and cuttings of seed plants. These asexual methods of multiplication are, on the whole, more efficient than sexual reproduction; farmers and gardeners often prefer them, and they have used them so long that some cultivated plants, such as the banana and sugarcane, produce only sterile flowers and can be propagated in no other way. Sexual reproduction, on the other hand, is wasteful. The sexual cells, male and female, are often produced in huge numbers, most of which, failing to find appropriate partners, die without issue. Sexual reproduction must bring some great advantage to offset its extravagant indirectness, and this is its ability to shuffle around the genes that determine heredity until favorable combinations occur.

Long before the discovery of the double helix of DNA, the name "gene" was given to the hypothetical bearers of heredity. Each gene was supposed to determine some particular character, such as the stature of the pea plants or the color of their seeds in the famous breeding experiments of the Austrian monk Gregor Mendel, who laid the foundation of modern genetics. We now know that one gene may modify the effects of another, that several may be jointly responsible for a single character, and that the total constellation of genes in an organism may, in some measure, influence the expression of each. In all the higher organisms, the number of genes is great. The fruit fly, *Drosophila melanogaster*, which in the laboratory has contributed more than any other organism to our understanding of the mechanisms of heredity, has an estimated five to ten thousand genes. Man doubtless has many more. Each gene appears to be a definite part of the long skeins of DNA molecules that constitute a chromosome.

The chromosomes are darkly staining, usually rodlike bodies that, under high magnification, are visible in the nucleus of a properly prepared cell that is about to divide. The cells that compose the bodies of most higher plants and animals have a double complement of chromosomes, half of which were contributed by each parent. The actual number is highly variable: the fruit fly has 8 chromosomes in each of its body cells; man has 46; some plants have well over 100; while the record appears to be held by the black mulberry, with 308 chromosomes in its vegetative cells.

In ordinary cell division, such as occurs in the growth of tissues, each chromosome splits lengthwise, and the missing part of each half is promptly replaced, so that the nucleus of each daughter cell contains a set of chromosomes identical with that of the parent cell. Unless a mutation happens to occur in some body gene, all the millions of cells in a large animal or plant will have exactly the same genetic constitution. This is why, when a grower wishes to insure that some excellent fruit tree or exceptional flower will remain true to type, he depends upon asexual propagation, which makes use of vegetative cells.

In the formation of the sexual reproductive cells, the sperms and eggs, nuclear division follows a quite different course. Now corresponding chromosomes line up in pairs on the equatorial plane of the nucleus, and one whole member of each pair goes to each daughter nucleus, which, accordingly, has half as many chromosomes as the body cells. When, in the formation of spermatozoa by a male animal, the paired chromosomes separate, all those from the animal's father do not go to one pole, while all those from his mother go to the other, resulting in only two kinds of sperm cells. On the contrary, the separation of the chromosomes is random, so that each spermatozoon may have any possible combination of the paternal and maternal chromosomes. A similar random distribution of chromosomes occurs in the formation of ova by the female. Thus, except in the improbable case that the two chromosomes of every pair are genetically identical, a very great diversity of sperms and ova will result. When the 46 chromosomes in the human body are reduced to 23 in the spermatozoa, over 8 million different combinations are possible; and the number of possible types of ova is equally great.

Since in fertilization any one of these diverse spermatozoa may enter any one of the ova, the number of possible combinations of chromosomes in the newly developing individual is a truly astronomical figure. With such tremendous opportunities for diversity, it would be little short of miraculous to find two people quite alike, except identical twins, who originate by the fission of a single fertilized egg and, accordingly, have just the same chromosomes.

As though randomly occurring mutations plus the random shuffling of the chromosomes at the formation of the reproductive cells did not permit sufficient opportunity for variation among the individuals of an interbreeding population, geneticists have discovered yet another source of diversity. When the chromosomes join in pairs in preparation for the nuclear division that reduces their number, the closely intertwined partners may exchange corresponding sections of their length, along with the genes that these sections bear, a process known as crossing over. When these chromosomes separate, each contains the same number of genes, but they may have different potentialities: a gene for blue eyes may be exchanged for a gene for brown eyes, for example. In other cases, a length of one chromosome breaks away and becomes attached to its partner, so that, after separation, one is shorter and the other longer; and this may be repeated until one chromosome permanently disappears from the lineage. In yet other cases, part of a chromosome becomes inverted, without losing its original position in the rod, and the resulting alteration in the arrangement of the genes may influence their effect upon the new generation. Thus we have mutations of two kinds: gene mutations, or alterations in the molecular structure of the genes, and chromosome mutations, or changes in the origin, number, or order of genes in a chromosome.

Chance enters into sexual reproduction as into a game of cards. At fertilization or conception, each future player in the game of life receives two sets of cards, one set representing the chromosomes it receives from its father, the other the chromosomes it receives from its mother. Each card bears a large number of symbols, which represent the genes. The symbols on corresponding cards may be identical, but often they will differ more or less; for example, the symbol for hair color may be black on the maternal card and yellow on the paternal card. After fertilization, the two sets of cards are mixed together, and copies of the whole double set are distributed to every cell of the growing body.

When the animal matures and is ready to reproduce, a single set of cards must be supplied to each of its germ cells. This set is prepared by joining in pairs corresponding cards from the two sets that it has carried all its life, then taking one card from each pair to make the new set; but from each pair it may take at random either the paternal card or the maternal card. In a similarly random way, the single set contributed by the other parent to the new life is made up. Accordingly, the double set composing the hand of cards received by the progeny may differ from that of either of its parents, any of its grandparents, and indeed any of its ancestors however remote, as well as from that of all its brothers and sisters.

To make this genetical card game more complicated and exciting, when the cards are laid in pairs preparatory to dealing them out to the germ cells, two cards may exchange symbols that have different values. Or part of one card may stick to another and be torn away from the first. Or the number of cards in the deck may be permanently increased by splitting each into two layers, so that even the single set in a germ cell will contain duplicates—a phenomenon known as polyploidy, which occurs more frequently in plants than in animals.

Life might be said to have gambled its way from its humble beginnings to its present tremendous diversity of forms, some still fairly simple, others highly developed. By a procedure that closely resembles a game of chance, each new individual plant or animal is given the set of genes that determine what it will be. If it receives a fortunate set, composed of sound genes all compatible with each other, it may flourish, living long and leaving progeny that will continue to swell the stream of life for ages. If the set is in any way defective, the organism will languish and finally succumb in the struggle to exist; indeed, if the set is most unlucky, containing a lethal gene, it may not live beyond the embryonic stage.

Natural Selection

Many kinds of plants and animals reproduce at such a rate that the world would soon be unable to support their progeny if all survived. As a result of mutations and genetic shuffling, individuals of a species vary more or less in their competence to meet life's demands. Since some reduction in their numbers is inevitable, the more fit will, on the average, survive better than the less fit. The weeding process has, rather unfortunately, been designated "natural selection," a term suggested by the breeder of plants or animals who selects, to receive special attention, the individuals that most closely approach his ideal of beauty or utility. But in nature we find nothing of this sort, no selection with any ideal or goal, no special care given to individuals that show some marked superiority; on the contrary, by their own unaided efforts, they must escape the hazards that destroy their less well endowed contemporaries. "Natural elimination" would be a more appropriate designation for what occurs. More accurately, what we witness is differential survival and reproduction. The stock or genotype that, in the long run, leaves most descendants wins the race to survive.

The agents of selection or elimination include all those circumstances that strongly affect vegetable and animal life. In time of drought, the individual with a more efficient water economy may survive while its neighbors perish. During great cold, the animal with thicker fur or other insulation, a more efficient metabolism, or the ability to reduce its metabolic demands by hibernating in a sheltered spot will be most likely to survive. In impoverished soil, a plant that needs a minimum of some scarce mineral element may thrive while more exacting individuals languish. When food is scarce, the animal with the most efficient hunting technique has the best chance of avoiding starvation. In extreme circumstances of any kind, quite small genetic differences, such as might occur among siblings, may spell the difference between life and death. Nevertheless, it would be wrong to conclude that the race to survive is always won by the strongest, the fittest, the most enduring. Chance enters here no less than in the distribution of genes: a fortunate accident may save a weaker creature while its stronger brother perishes.

Although the lifeless environment eliminates many organisms, the chief agents of destruction are living things themselves. Among plants, competition is sharp, especially among crowded seedlings. It occurs not only in the soil, where roots compete for mineral elements and water, but even more above the ground, where shoots race upward and the most rapid grower overshadows its neighbors, depriving them of essential sunlight. Nonparasitic plants do not actually attack each other, although vines that spread a smothering blanket of foliage over shrubs and even over the crowns of the tallest trees of tropical forests might be considered to be actively aggressive.

In the animal kingdom, strife is far fiercer and more direct. In an overcrowded world, where survival often depends upon the ability to overcome adversaries or capture prey, a method of evolution capable of altering almost every organ in every conceivable manner was bound to produce a formidable array of offensive and defensive weapons: claws and talons for seizing, fangs and beaks for tearing, horns and sharp hoofs for attacking and defending, venomous stings for inactivating prey or deterring aggressors, and an amazing variety of snares and traps for capturing victims. This mode of evolution promoted the swiftness of the hunter as well as the fleetness of the pursued; every increase in the one was necessarily followed by an increase in the other, for this was the condition of survival for both the prey animal and the predator that depended on it for nourishment. And, since psychic traits are no less mutable than physical characters, along with weapons evolution promoted the fierce aggressiveness indispensable for their effective use. In this harsh school the human stock, which for some millions of years has been predatory, developed the belligerence, callousness, and cruelty that are our torment and our shame.

In a crowded world, strife arises not only between species but within species. Intraspecific competition is, in one respect, the sharpest, because members of the same species have the same needs, while the needs of different species are more or less different. But, except in social species that make war, including man and such hymenoptera as bees and ants, intraspecific competition tends to be more subtle, less violent and spectacular, than interspecific predation.

Yet in one particular context, where we would least expect it, we often witness violent strife between members of the same species—the struggle of males for females. Sexual reproduction requires fine adjustments: the male and female must be simultaneously ready, perhaps in one short interval each year; they must be able to excite positive responses in each other; the reproductive cells supplied by the two sexes must be so similar biochemically that they fuse into one; and their complements of genes must be sufficiently alike to promote harmonious development of the resulting embryo. How could strife enter this realm of delicate adjustments?

The answer seems to be that the vigor, fierceness, superior armament, or whatever enables one animal to overcome sexual rivals and win the greatest number of females or to hold the most attractive and productive breeding territory will, transmitted to his progeny, aid them in the strenuous business of living. Sexual fighting appears to be a byproduct or a secondary development of the general struggle to survive in a harshly competitive world. In a peaceful community of animals subject to no great environmental stress, it would have much less survival value and doubtless be less frequent and fierce. And it is but fair to add that among many animals, especially birds, sexual rivalry has become formal and nonviolent. The males vie to win mates by their often profuse adornments and elaborate displays, without clashing together; the female freely chooses, to father her offspring, the performer who most attracts her. The male's vitality is revealed by the splendor of his plumage, the vigor and persistence of his performance, rather than by his prowess in battle.

Scientists continue to argue whether natural selection acts only upon individuals or also upon groups of organisms. Individual selection, over many generations, seems adequate to account for the evolution of all those characters—anatomical, physiological, and behavioral—on which the survival of individuals, by their own unaided efforts, depends. Strength, fleetness, dentition, metabolic efficiency, protective or warning coloration, potential longevity, and many other characters fall into this category. But individual selection fails utterly to account for social interactions and all the structures and functions that support such interactions, which are devoid of utility or selective advantage, which by definition cannot exist, unless they involve at least two individuals.

For social interactions to evolve, associated individuals must vary in a coordinated manner and be selected together. Signal and response, releaser and innate releasing mechanism, must, through their evolution, correspond like key and lock. A cry that warns of danger would be a meaningless sound without some other animal that responded to it. If the nuptial attire or courtship antics of a male bird changed without a corresponding alteration in the preferences of a female of his species, that male would leave no descendants. The existence of social interactions of all kinds points unmistakably to the selection of groups of interacting individuals.

Although fervid individual selectionists deny that group selection is effective among populations of interbreeding individuals or demes, the group molded by natural selection may transcend the limits of species. In tropical seas, a number of small fishes nourish themselves with parasites and dead tissues that they remove from larger fishes who actively seek their services. How could the association of these cleaner fishes and their clients arise without the simultaneous modification of the habits and even structures of an interspecific group consisting of both the cleaners and the fishes that solicit their attention? Indeed, the species selected together may be far more dissimilar than two kinds of fishes; they may even consist of an animal and a plant. Consider, for example, the coevolution of ants and the bull-horn acacias which they inhabit in Mexico and Central America. These small trees offer the ants homes in paired hollow thorns and nourish them with tiny protein corpuscles produced at the tips of their diminutive leaflets. In return for this generous hospitality, the ants protect the acacia trees from insects, browsing animals, and encroaching vegetation. Ant and plant have been reciprocally adapted by natural selection acting upon both simultaneously for generations. Dogmatic insistence that only individual selection is effective in evolution springs from focusing attention upon individual genes, rather than whole complexes of genes, and upon competition among organisms while overlooking the many forms of cooperation that the natural world presents. To deny the reality of group selection is to deny that social interactions influence the course of evolution, which is absurd.

Just as group selection may foster helpful social interactions, so it may prevent the spread of injurious behavior. We humans hardly need to be reminded how easy it is for an individual to multiply his progeny at the expense of his neighbors. Among birds, this is more difficult, yet we can imagine certain behavior by which a bird might increase its reproductive output at its neighbors' expense. A colonial bird might solicit so compellingly that parents bringing food to their young would relinquish it to the beggar, who would then deliver it to its own nestlings. Since cannibalism is by no means absent from colonies of gulls, frigatebirds, and others, a parent might nourish its brood with chicks from neighboring nests. Unscrupulous parents would raise more progeny than their well-behaved neighbors, with the result that the colony would become increasingly infected by their vices. As the number of beggars or cannibals grew, such misconduct would become less profitable, and it might fall heavily upon their own offspring. The population in which it became rife might dwindle and vanish, to be replaced by other groups with sounder habits. However it is achieved, excessive reproduction in a season of abundance can be harmful, causing a drastic decline of the population by starvation during the following period of scarcity.

When we take the short view, it is obvious that individuals are selected. Some live longer and leave more progeny than others, hence are fitter in the evolutionary sense. Nevertheless, in sexual reproduction, the constellation of genes responsible for this fitness cannot pass unaltered to the next generation, unless the individual finds a sexual partner with an exactly similar set of genes (except, of course, those which determine sex). In their offspring, the combined contributions of both parents are tested together for compatibility and fitness. In the third generation, the contributions of four grandparents, joined in different combinations in different grandchildren, are screened by natural selection—and so, increasingly, as the generations pass. Although in the short view individuals are selected, for elimination or preservation, in the long view groups of interbreeding individuals are selected. If the group survives, what ultimately emerges from the continual shuffling of genes and the testing of each new combination in individuals is a population with a self-perpetuating pattern of life. For the immense variety of patterns, even among species in the same family, the chances of mutation and recombination, together with the great variety of available niches or habitats, are responsible.

Recently, naturalists have given much attention to the concept of kin selection. Next to leaving direct descendants, an animal can best multiply its genes by helping its closest relations, parents and siblings, rear their young, as in cooperative breeding by birds. Kin selection helps us understand the evolution of heroic self-sacrifice in defense of the family or clan. The kindred saved by the hero share many of its genes, probably including those responsible for its bravery, so that the tendency to risk life in defense of others will not be extinguished. But all the members of an interbreeding natural population have many genes in common, as is evident from their similarity in structure and behavior. Kin selection merges into group selection. The selfish individual who does nothing for the perpetuation of its species helps prepare the way for the extinction of its own genes, for only in a species can they survive for generations. Even when they compete for success in reproduction, individuals cooperate to perpetuate their species.

Individual selection in its purest form occurs in a population derived by vegetative or asexual reproduction from a single parent. Such a population, known as a clone, consists of genetically identical individuals, which are, in effect, detached parts of their progenitor. Plants, which multiply by fission, bulbils, tubers, detached branches, or buds, readily form clones, some of which, of importance to man, consist of immense numbers of individuals spread widely over our planet. Each named variety of a vegetatively propagated ornamental or economically important plant frequently consists of a single clone, whose origin can often be traced to hybridization or to a bud variation in somebody's plantation, garden, or greenhouse.

If the selection of the genotypes of individuals, unmixed with those of other individuals, promoted evolution better than group selection, evolution would doubtless have made asexual reproduction far more widespread than we find it. Although asexual or vegetative reproduction may preserve a successful type for many generations, especially under man's protection, it does not provide the genetic variability needed to confront changing habitats. Sexual reproduction offers a much greater range of genotypes to the action of selection. Hardly less important is the increased opportunities for the interaction of individuals that it affords. Clones can compete for space, nutriments, or propagation by man, but their opportunities for cooperation appear to be limited to the preservation of their habitat. Organisms that reproduce sexually cooperate more fully, by mingling genes from various sources to produce new and often superior forms and, if they are animals, by rearing offspring and forming societies. Likewise, they compete in ways that clones cannot, notably, for sexual partners. This possibility for closer cooperation along with intensified competition has been responsible for some of the most splendid, as well as some of the most deplorable, developments in the living world. We owe to it not only many admirable animal societies but likewise our own morality and altruism.

Harmonization and Growth

How shall we regard this process of evolution that has filled the seas and covered the land with varied life, this movement to which we and all our fellow creatures owe our existence? What attitude shall we take toward it? Certainly, it is most important for us to have a correct attitude, for our views of nature and of man, our attachment to our sources and our assessment of our prospects, will be profoundly affected by our appraisal of the process that shaped us. Many scientists who have a deep understanding of the causes and course of evolution seem never to have defined their attitude toward it, or they seem to have what appears to be an incorrect attitude. They cannot all be right, for these attitudes range all the way from that of Darwin's staunch supporter, Thomas Henry Huxley, who regarded evolution as a cruel, evil process that the moral man should stoutly resist, to that of his erudite grandson, Sir Julian Huxley, who proposed evolutionary humanism as a developed religion.[1]

Evolution begins with accidents in the form of fortuitous alterations of the molecules that determine heredity; it proceeds by a method that closely resembles human gambling; it consolidates its gains by ruthlessly destroying the losers, however excellent they may be in themselves, if they do not fit well into their actual environment or cannot meet the stress and competition to which they are subjected. By creating wonderfully efficient organs, such as the eyes and ears of birds, and creatures marvelously adapted to their environment, it gives an impression of superhuman efficiency. The impression is false: to accomplish these things, evolution blundered ineptly for millions of generations,

[1] T. H. Huxley and Julian Huxley, *Evolution and Ethics 1893–1943* (London: Pilot Press, 1947).

hoarding its accidental gains and obliterating its far more numerous failures, like a bookkeeper who enters only his credits in his ledger. What is there in all this harsh, mindless business to win our admiration or stir our loyalty? A reverent regard for the sources of our being ennobles and stabilizes life; but who can cultivate piety toward accidents, gambling, and ruthless destruction?

Moreover, neither accidents, nor gambling, nor the wholesale elimination miscalled natural selection is constructive, yet, when we look around us and contemplate what evolution has produced in the living world, we cannot doubt that construction has occurred. Evidently something has been omitted from the foregoing account of evolution, which is, in broad outline, that generally accepted by contemporary evolutionists, with the omission of many details that the interested reader can find in hundreds of textbooks and special treatises. On the whole, these accounts at best imply a constructive principle that they fail to specify. Some go so far as to say that natural selection is constructive; but this, as has become evident, is above all the destruction of the types less well fitted to survive, however admirable they may otherwise be. And, unless construction had already occurred, there would be no organized beings to destroy.

The driving force in evolution is the creative energy that pervades the Universe, revealing itself as harmonization, the movement that builds its primitive materials into integrated patterns which realize some of the values latent in it. One of the most revealing examples of this movement is the growth of a green plant, in which simple substances, dispersed in air and soil and water, are absorbed and built— molecule by molecule, cell by cell, organ by organ—into a coherent, harmoniously functioning whole, which is not only beautiful but carries on the beneficent labor of photosynthesis.

The growth of a plant is a purer expression of harmonization than that of an animal, which must destroy other living things to procure materials for building its own tissues.

Nevertheless, animals, at least those of more advanced types, carry harmonization to higher levels than do plants, for their organs are more complex, more varied, and more closely integrated by means of a rapid circulation and a nervous system reporting to a central sensorium. Moreover, they feel and think as plants evidently cannot do; they sometimes form societies that cooperate in ways more varied than is possible for vegetable associations; and, although unable to synthesize the stuff of life from simple inorganic materials as plants can, some of them create excellent things in manifold ways.

Throughout the long course of evolution, from the first alterations of the most primitive forerunners of life to its latest creations, the most constantly operative process has been growth. The evolutionary processes of mutation, sexual shuffling, and selection are, above all, methods of originating and testing different patterns of growth. Mutations change the rate and direction of growth; permutations of genes in sexual reproduction cause further variability in growth; natural selection makes room for certain kinds of growth by eliminating others. The evolution of life on this planet is, above all, the story of changes in the rate, direction, and form of growth. Without growth there could be no evolution. And growth is a mode of harmonization, the creative movement that pervades the Universe. This process that forms us in body and mind, making each of us what he is, can win our unstinted approval and loyalty as evolution fails to do.

Many of the less pleasant aspects of evolution, the strife and the carnage, may be attributed to the excessive intensity of the organizing movement, which creates so many living things that they inevitably compete with each other for the space and materials that they need for growth and, in the absence of other sources of nourishment, devour or parasitize each other. Such natural catastrophes as volcanic eruptions, earthquakes, floods, hurricanes, and droughts severely afflict living things, including man, but in the aggregate they are responsible for a quite minor share of the ills that the living world suffers. Whereas these excesses of nature are local and intermittent, often sparing great expanses of Earth for long periods, the conflict of living thing with living thing, in the form of predation, parasitism, infectious disease, and the less violent competition of plants for a place in the Sun, continues day by day and hour by hour wherever life abounds.

If life were not so superabundant, living would be much more pleasant, relieved of a large share of the fears, pains, discomforts, and annoyances that now afflict it. A far larger proportion of the animals that are born, of the plants that germinate, would grow to perfection and live their full term. The excessive fecundity of organisms is at once the triumph and the tragedy of the living world; it insures that every niche capable of supporting some form of life, however humble, will eventually be occupied; but the price of this ubiquity is strife and suffering. Yet possibly, if life were not so excessively prolific, evolution would be slower than it has been, requiring even more than the several thousand million years that it has taken to rise from its earliest beginnings to ourselves.

As we have seen in chapter 2, most of the matter in the Universe exists in conditions that restrain its creative powers. Much lies crushed and suppressed in the interior of massive celestial bodies; much more is too thinly diffused in interstellar space; some is too hot and some too cold. When, in rare favorable circumstances, such as on the surface of our planet, matter can enter into the complex living state, it displays excessive exuberance, like superheated water suddenly released from pressure or like schoolboys rushing out after a long session in the classroom. It starts far more living things than Earth can support, suggesting that the goal of all matter may be the living state, which most fully expresses its Daedalian skill in creation—but at the price of tragic conflicts.

Life, organization, is the source of most of our values and

must be accounted good. Evil, which is the destruction of values or their replacement by such disvalues as ugliness, falsehood, and pain, springs from the very intensity of the movement to bring good into the world, rather than from the action of some principle of evil, some wicked spirit, as certain thinkers have supposed. If the movement to bring good into the world were less intense, this world would be less evil. Evil is the secondary effect or by-product of the very process that creates good, especially when this process becomes excessive—an alarming but inescapable conclusion from our survey.

Evolution reminds us of animals trying to find their way through a maze with rewarding food at its exit, such as students of animal behavior use to test the ability to learn. This labyrinth has many turns, many culs-de-sac or dead ends. From the latter there is no returning; whole species, families, and orders of animals and plants have blundered into them and been utterly annihilated, as happened to the seed ferns that flourished in Paleozoic times and to the giant saurians of the Mesozoic. Contemporary plants and animals, including man, are still successfully following the maze, finding food and multiplying as they go. The exit will be reached, if at all, by the being that most fully realizes life's creative power, so that no further advance is possible. Whether this creature that has triumphantly traversed the perilous labyrinth will be man or his remote descendants, or a development of some other stock, remains to be seen.

How much time might have been saved, how much suffering avoided, if life could have been guided through the labyrinth by a wise, compassionate mind! For aeons, the matter of the Universe has been adequate for the formation of the most admirable creatures that we know. What has been lacking has been a being with the knowledge, the power, and the will to build up the atoms, promptly and swiftly, into the most beautiful, the most joyous, the most loving living things that they could form, all dwelling together in concord. In a manner of which we are wholly ignorant, the Universe acquired the capacity to bring forth the most precious things, but, as far as we can see, it has had to realize its potentialities by the painful, time-consuming process of trial and error. What stirs our wonder is not so much its blunders as the persistence of its efforts, which suggests that what can be will be, no matter how many ages it takes to achieve this.

The world in its present state reminds us of an early model of some complex machine, the first workable automobile or airplane or typewriter, which operates with many flaws and frequent failures. Yet it has great possibilities and with time and patience might correct its errors. But where is the inventor trying intelligently to improve the world machine, unless it be a few of us humans with more goodwill than wisdom and power?

Even a cursory survey of evolution reveals the tragic paradox of our planet, perhaps of the whole Universe. It has such splendid potentialities, but to realize them is so difficult. When, to all the marvelous structures and capacities of the living world, we add all that technological man has achieved, mostly in the last two centuries, we suspect that the materials which compose the planet and are widespread in the Universe can be made to do almost anything we can reasonably ask of them. They have an almost unlimited capacity for creating all that is good and beautiful, as well as much that is evil and ugly. And, when we reflect that all creative advance, in the living no less than in the technological world, depends largely upon the patterns in which trillions of only three basic particles—protons, electrons, and neutrons—are arranged, this almost infinite capacity appears the more wonderful.

In recent times, chemists have learned much about the arrangement of atoms in molecules, and with this understanding has come some ability to transform molecules for specific purposes, although they are still unable to synthesize the simplest living thing. What might have been done with them by a superchemist, determined to make the best of these so versatile and pliant materials, while avoiding all unfortunate combinations! In the absence of such benevolent supervision, life has had to blunder its way upward by trial and error, dependent upon chance recombinations of atoms for each advance. Evolution has been a slow and painful process; its achievements have been dearly bought with countless lives and immeasurable suffering. Although in certain moods we may doubt whether it has been worth the price, when we view our planet at its smiling best, doubts vanish. But perhaps we should withhold final judgment until we see what man, who has gained a considerable ability to influence evolution, in himself and, at least indirectly, in the whole living world, finally makes of it.

Any organized effort to guide the course of evolution would confront a moral problem that a survey of evolution raises in a thoughtful mind. That evolution has accomplished much that is splendid and admirable, it would be ungrateful to deny. That the means it has employed have often been ruthlessly harsh is a proposition to which every compassionate person will assent. Yet, as far as we can see, no other procedure for raising life to higher levels was available, and without these crude methods life would have continued until today at a primitive level. Can we conclude from this that ends, if sufficiently grand, justify means, however horrible? No! The course followed by a blind process can afford no precedent or justification for the actions of a rational, moral being. Evolution has raised us to a moral level that forbids us to follow its example, and this is one of its finest accomplishments. Moreover, the end bears tragic flaws that must be attributed to the crudity of the means.

8. Man as a Product of Evolution

 Men commonly regard themselves as the highest product of evolution on this planet, the most excellent of its teeming inhabitants. Although this claim might be contested by other creatures, it is objectively certain that man is one of the latest of evolution's creations and likewise one of the most advanced, in distance measured from life's beginning. Let us glance briefly at what evolution has made of us, in body and mind.

The Human Body

By almost any standard, the human body is one of the best, if not the absolute best, in the whole animal kingdom. Like that of other homeothermic animals, the mammals and birds, it maintains, over a fairly wide range of ambient temperatures, a constant body temperature and capacity for activity. Although it cannot, like polar bears, penguins, and other animals of high latitudes, endure great cold without artificial protection, it is quite adaptable to environmental extremes; few animals can flourish in such varied climates.

Even without artificial aids, man is capable of more varied activities than any other animal. He can walk, run, climb, swim, and dive. To be sure, he cannot compete with the specialists in any of these activities, running as fast as gazelles, climbing as well as monkeys, swimming and diving as well as dolphins; but in locomotor versatility he excels them all. His supple hands are the most serviceable manipulatory organs in the whole animal kingdom, capable of performing more varied operations, of making a greater diversity of things, than the limbs of any other creature. And, with these so versatile hands, man can make machines that enable him to outrun the gazelle, cruise the seas more swiftly than the dolphin, dive deeper than the whale, and even fly faster and higher than any bird.

With sense organs we are well equipped. Our vision is less keen than that of many birds, our hearing less acute than that of many birds and mammals, our sense of smell decidedly less sensitive and discriminating than that of many mammals. On the other hand, our naked skin and sensitive fingertips make us superior in the sense of touch. Color vision, which we share with other primates, birds, and at least some insects, doubtless makes our visual discrimination superior to that of the many mammals that lack this refinement, and it certainly adds immensely to our enjoyment. Again, although certain animals have one or two senses that are better than our own, taken as a whole our sensory equipment seems as adequate as that of any other creature, if not superior. This, as we shall see, has far-reaching consequences for our role in the Universe.

In our ability to produce varied sounds, we have no rivals except mockingbirds and certain other avian mimics. For all their vocal facility, birds have at best rudimentary speech; they can express emotions and convey signals but are unable to form sentences and communicate concepts, a limitation imposed by their mental rather than their vocal development. Another great advantage that we enjoy is our ability to digest, and nourish ourselves with, the most varied foods. This gastric versatility not only permits us to survive on the local products of the most diverse climates but also enables us to select our diet with regard to moral and economic considerations, a freedom that would be denied us if our choice of foods were as narrowly limited, by physiology, as that of certain other animals.

Along with these great advantages, our bodies inherit certain defects. Our vermiform appendix, a vestige of a diverticulum that was doubtless useful to remote ancestors who consumed large quantities of foods of low nutritive value, is frequently infected, with, until the advent of modern surgery, fatal consequences. Probably appendicitis is, in the long history of the human stock, a fairly recent disorder; otherwise, we would expect the appendix to have been eliminated by natural selection. Our upright posture, subjecting our abdominal wall to strains that are more uniformly distributed in quadrupeds whose viscera are slung below the backbone, makes us too liable to rupture. Our wisdom teeth, useful to big-jawed ancestors who munched tough foods, often become impacted and cause trouble. The rest of our teeth, which we need, are lamentably prone to decay; but this seems to be largely the fault of improper care, along with a diet too soft and sweet. We are, of course, subject to many diseases, both infectious and degenerative; but other animals are by no means exempt from such afflictions; and we, as they, can produce antibodies in our blood to give us immunity to infections.

Despite these weaknesses and defects, our body is remarkably tough and enduring. Few other animals, larger or smaller, are so long-lived as man. Although man's average life span has been greatly prolonged by technological civilization, his potential longevity, determined by innate factors, has probably changed little since Neolithic times. Even with protection and the best of care, domesticated animals and those in zoos rarely live as long as man, while free animals are beset by so many hazards that they seldom die of advanced age. Among mammals, only elephants and some great whales appear to have a potential life span equal to man's. In the tough game of evolution, we have accumulated some of the best cards, bearing the genes that give us exceptionally versatile bodies, well equipped with sense organs and capable of enduring a long while. It is hard to understand how people who have had the excellent fortune to receive such bodies can so often gratuitously abuse them, with injurious foods, drinks, and drugs, with inadequate rest or exercise, and in countless other ways.

Mental Qualities

Man excels other animals, above all, in mental qualities, among which we may include reason, imagination, sympathy, aesthetic sensibility, and similar attributes. Although the exact relationship of mind and brain is still obscure, the two are somehow related; in the long history of the human stock, from the earliest hominids to modern man, advances in social organization, practical competence, and intellectual ability have been accompanied by increasing cranial capacity. Modern man has a bigger brain than any other animals except elephants and whales, which are many times heavier. It is certain that this brain did not originate as an organ of pure reason, which even in the majority of our contemporaries is poorly developed. The growth of the human brain was promoted by the practical advantages it gave to its possessor in the struggle to live; and its history is so intimately connected with man's upright posture, versatile hands, speech, and sociality that, to be understood, all must be considered together.

Man has been called both the rational animal and the tool-using animal, but the two faculties are so interdependent that it would be more correct to characterize him as the rational, tool-using animal. Although this conjunction would have been vehemently rejected by Greek philosophers, who disdained manual labor and prided themselves on their rationality, it is certain that if man were not one of these things he would not be the other. According to David Pilbeam, "It is only long after stone tool–making traditions became established that we find the human brain growing to its present relatively great size."[1] Our reason is no divine gift but an incomparably precious prize won in the long evolutionary gamble.

It is sobering to reflect upon the momentous consequences of apparently insignificant gains in this protracted

[1] David Pilbeam, *The Evolution of Man* (London: Thames and Hudson, 1970).

game of chance. If our remote prosimian ancestors, who adopted an arboreal life some 70 million years ago, had climbed by digging sharp claws into bark in the manner of squirrels and many other scansorial animals, instead of grasping branches with fingers and toes, as modern monkeys do, I would not be holding a pen in my hand as I write this. During the long ages when our ancestors dwelt in trees, not only were their hands and feet improved as grasping organs, but they came to depend more and more upon sight rather than upon scent, since only vision could direct the tremendous leaps they often made while scurrying through the treetops. When, in an era of increasing drought, forests shrank and the primates destined to become our ancestors were forced to live on the ground, they had grasping hands, probably some ability to walk upright as apes can do today, and well-developed vision with forwardly directed eyes. Their brains were hardly larger than those of contemporary great apes.

With this equipment, which they owed largely to their long life in trees, primitive hominids began their hazardous march toward full humanity. The goal would never have been reached unless the four correlated faculties of upright posture, manual dexterity, intelligence, and speech had kept pace with each other. The value of versatile hands would be greatly diminished if they had also to act as feet for locomotion. Released from this service, what these hands could accomplish depended upon the mind that guided them. Mind itself, no matter how brilliant, could contribute little to survival without organs to put its valuable ideas into effect. Some of the projects, even of subhuman hominids, were doubtless so big that they could be realized only by group cooperation, and this is where verbal communication acquired survival value. As skills slowly accumulated and improved and social organization passed beyond the purely instinctive level, language was needed also for instructing the young and transmitting tribal lore from generation to generation.

In retrospect, upright posture, manual skill, brain size with associated intelligence, and language appear to have been bestowed upon the human stock in a package. Nevertheless, each advance—in posture, structure of hand and forearm, size and quality of brain, and linguistic ability— must have depended upon separate favorable mutations, among a much larger number of mutations that were useless or wrongly directed. The marvel is that so great a transformation as that from the first tool-using hominids to *Homo sapiens* was accomplished in only about 2 million years, a short stretch of geological time.

Those who write about the evolution of man frequently pass too lightly over all that we owe to our remote arboreal ancestors and the trees that supported them, while they emphasize the importance of the cooperative hunting of large animals in shaping human character and social life. They remind us that for an immensely long period pre-agricultural man and his ancestors hunted in packs, like wolves and certain other carnivores, and that to plan and effectively execute the hunt required forethought, leadership, communication, coordinated movements of all the hunters, and mutual trust in perilous situations. But, when we recall that wolves, hunting dogs, and other quadrupeds successfully attack and bring down animals much bigger than themselves without developing large brains, high intelligence, language, or any society more highly organized than the wolf pack, we may doubt that hunting in packs has contributed importantly to human evolution. It was probably a means of supplementing, especially in time of scarcity, a diet that was still largely vegetarian, as well as helping sustain the human lineage until, about eight or ten thousand years ago, man developed agriculture.

More important than the hunt itself in developing human minds and hands was the manufacture of its weapons: hand axes, or spear tips of chipped flints, or whatever else they used. This was a constructive activity that demanded precise movements of hands guided by minds with an image of the article they were shaping—an activity more apt to promote the development of brain and hand than the wild, barbaric excitement of the chase. Hunting by man's ancestors appears to have received disproportionate attention because its instruments of enduring stone and its spoils in the form of cracked skulls and shattered bones have persisted in fossil deposits, while other items that these same people doubtless made of vegetable materials decayed long ago. The latter may have included rude shelters, because caves and overhanging ledges of rock are lacking in many otherwise suitable territories.

The accounts that dwell upon cooperative hunting by males, especially youths, say too little about females and their influence upon man's evolution. Since females transmit to posterity just as many genes as males do, should they not contribute about equally to its shaping? As brains grew bigger and babies were born too helpless to cling to their mothers' increasingly hairless bodies, as infant monkeys do to their mothers' fur, the females' need of hands to carry their babies may have been a powerful factor in the perfection of the bipedal locomotion that left forelimbs free for other uses.

The division of labor among recent hunter-gatherers, among whom men hunt while women gather edible fruits, nuts, tubers, and other vegetable products, appears to be the continuation of a practice that began ages ago, possibly among prehuman hominids. Just as the males hunted in bands, so the females searched the surrounding bush in chattering, cooperative parties, while they taught their daughters how to recognize edible plants and avoid poisonous ones and how to prepare each desirable species. It would be surprising if, at a period when males fashioned weapons of stone, females did not make baskets or bags that would so greatly help them carry their harvest to their camp. Certainly neither manual dexterity nor intelligence has ever been restricted to one sex. While the males who hunted large animals, developed weapons, and inured themselves to slaughter finally became warriors who slew their fellow men in ever more sanguinary wars, the females who tenderly nurtured young increasingly dependent upon them contributed a softer, more cherishing, more amiable side to developing human nature.

After brain size and intelligence reached a certain critical level, this led the way in human evolution, which from being biological gradually became cultural. Although these two modes of evolution are fundamentally different, they have a certain curious resemblance. Biological evolution depends upon random genic mutations and the elimination of those that fail to promote survival. Cultural evolution depends upon ideas that often spring up in the mind as unexpectedly and inexplicably as mutations in the body, upon their testing in practice, and upon the final rejection of those that prove unprofitable. Thought, like evolution, advances by the method of trial and error, by proposing answers to our practical problems or theoretical questions, examining them carefully, and dismissing them when inadequate. And, just as most genic mutations are worthless, so only a minority of the ideas or brilliant proposals of restless minds have permanent value.

In any case, advances in techniques, whether in chipping flints for stone axes or designing airplanes, as well as in living habits and social organization, depend upon mental activity, upon the free association of ideas. Already toward the end of the Ice Age men who made stone implements, clothed themselves in skins, and traced on the walls of their caves spirited drawings of the animals they hunted had skulls as capacious as those of modern man. Possibly their minds were of equal intrinsic quality, yet, in addition to some knowledge of the ways of the animals on which they preyed and the uses of the plants they gathered for food, as well as of the local topography, their heads were doubtless filled largely with absurd superstitions, irrational fears, and strong passions. Today, with brains no bigger but with the aid of a long tradition of rigorous intellectual discipline, we master elaborate sciences and develop the most intricate apparatuses. Man's mental capacity evidently raced ahead of his immediate needs; at the end of the Paleolithic period he was already preadapted for the complexities of technological civilization; and even today few of us seem to make the fullest use of our minds. Indeed, as we grow older, our cerebral neurons are said to waste away by thousands, probably because they remain unused, since this destruction precedes the mind's senile decay.

Some writers, including Alfred Russel Wallace, coauthor with Darwin of the theory of evolution by the natural selection of small, random variations, have concluded that the large brains of men still in the primitive state could not have been acquired in the usual course of evolution. Natural selection, Wallace claimed, cannot produce characters in excess of immediate utility, among which he included not only our ancestors' incompletely used mental capacity but also the largely naked skin of races still without adequate clothing, as well as the moral sense. From these developments, anticipating the needs of men in a more advanced civilization, he inferred that "a superior intelligence has guided the development of man in a definite direction, and for a special purpose, just as man guides the development of many animal and vegetable forms."[2]

We now know that mutation can produce characters in excess of actual utility, especially when these characters are linked with others of immediate value. Natural selection will not operate to repress excessive or useless characters unless they become an impediment, in which case they will be eliminated, often along with the species that has been burdened with them. The Irish elk was apparently dragged down to extinction by its enormously overgrown antlers. The possession of excess cerebral tissue is no evi-

[2] Alfred Russel Wallace, *Contributions to the Theory of Natural Selection*, 2d ed. (New York: Macmillan, 1871).

dent handicap, and it could even have been advantageous to savages who in their struggles with each other and with wild animals not infrequently suffered brain injuries, by permitting the development of alternative neural pathways. Moreover, the enlargement of the cerebral hemispheres is closely correlated with speech and with manual dexterity, of which primitive man had at least as much as his civilized descendants—he needed it to survive. Apparently, it was to control the hands that the human brain grew so large, and it only secondarily acquired the capacity to philosophize, to develop sciences, to create literature and art, as well as the other abilities for which many of us now chiefly value it. Yet even now it would be perilous to disregard the intimate connection between manual skill and cerebral development. As Haldane remarked, "If we bred for qualities which involved the loss of manual ability, we should be more likely to evolve back to the apes than up to the angels."[3]

Moreover, it is difficult to reconcile the belief that a superior—and benevolent!—intelligence has guided the development of man, with his many obvious imperfections. If this intelligence removed the body hair from shivering savages, in anticipation of the day when their descendants would wear tailored clothing and find hairless bodies easier to bathe, why did he not also suppress the hair on men's faces, where it is so easily soiled with food and so readily becomes unsightly unless daily shaved away? Why did he not correct the physical defects that were earlier mentioned, give men eyes that do not need glasses as they age, and make other improvements? Above all, why did he permit us to become burdened with the violent passions, so difficult to control, that we must now consider?

[3]J. B. S. Haldane, "Human Evolution: Past and Future," in G. L. Jepsen, E. Mayr, and G. G. Simpson, eds., *Genetics, Paleontology, and Evolution* (Princeton: Princeton University Press, 1949).

As sources of human suffering and shame, these corporeal imperfections, and others that might be mentioned, are of minor importance compared with man's psychic disorders, the evil passions that too frequently inflame his mind. Why are we so often aggressive, cruel, greedy, lustful, deceitful, fearful, and venomous with hatred? These traits that so distress everyone who aspires to goodness or holiness were evidently foisted upon humanity, as on other animals, in the long and often grueling struggle to survive in a crowded world. Mutations, as we have learned, may alter any of the characters of an animal's body and, through their somatic foundations, those of its mind. Thrown into conflict with each other for food, living space, and mates, animals acquired offensive and defensive armaments and the skills necessary for their effective use. As animals' psychic life developed, it was bound to correspond to the actual circumstances of their existence. The governments that in recent wars used every device of propaganda to stir up hatred of the enemy were only following nature's way, for hatred sharpens aggressiveness and the will to overcome. The predatory animal that ruthlessly tears the quivering flesh of its victim, if not already psychically cruel in the sense of gloating over the creature's agony, has all the elements of cruelty within it.

Man's growing capacity for thought and emotion only intensified the fear, hatred, vengefulness, greed, and deceit already implicit in the struggles of animals lower in the evolutionary scale. Competition has strengthened muscles, improved skills, and sharpened wits in man as in other animals; but the heavy price we have paid for these advantages is an array of violent passions that distress the individual, disrupt society, and incite wars between tribes and nations. Burdening our souls with these ugly passions was perhaps the most unkind thing that nature has done to us. To be so afflicted is the trauma of evolution, far more real and distressing than the so-called trauma of birth, which many of us cannot recall.

If our evolution had been guided by a benevolent intelligence, we would expect, at the very least, that when he endowed us with reason and forethought he would have given us full rational control over our reproductive powers. With the exception of those who direct the destinies of states and nations and those who make discoveries and inventions or elaborate ideas that transform our patterns of living or thinking, no man or woman does anything more momentous than begetting a new human life, with all its capacity for joy and sorrow, for glorious achievement or shameful failure. One would expect that nobody would undertake this awful responsibility without the most careful forethought, considering the genetic endowment that the baby would be likely to receive, the resources available for the child's nurture and education, the probability of his finding a satisfying occupation when he grows up. The only motive that could justify human reproduction is the generous desire, by one who has found living a highly rewarding experience, to give, as a free gift, a joyous and satisfying life to others. Even the prospect of domestic felicity amid a family of happily growing children, of being less alone when old, should be subordinate to this. One should approach parenthood with the hope of contributing to society a new member who would raise the quality of human life.

All too few are the children born in such auspicious circumstances. In and out of wedlock, millions are started on life's way with less careful thought than their parents give to buying a new garment or planning a meal. Driven by passions uncontrollably urgent, people beget children without preparation, without purpose, without the means to support them, without an ideal. The tragedies that result from casual parenthood, the wasted lives, the suffering from remorse and social reprobation, the bitterness of growing up without parental love and an approved place in society are visible on all sides and among the most frequent themes of literature. If insemination had been made painful and child-birth a pleasure, instead of the reverse situation that prevails, some of humanity's worst problems would have been avoided, and the quality of mankind would be immeasurably higher.

Many animals with an annual breeding season are sexually active for only a fraction of the year. During much of his life, man knows no such placid holiday from the mindless urge to beget children that he may not want. This situation probably did not arise until, by building shelters and insuring a fairly adequate food supply throughout the year, man had made all seasons almost equally favorable for raising young, as has happened also with his domestic animals. Even in societies with a primitive economy, human fecundity is often excessive, making it necessary to resort to abortion, infanticide, or restriction of sexual intercourse by tribal taboos, in order to hold the population within the limits imposed by the productivity of the group's territory. Even at an earlier stage, when hominids oppressed by fierce carnivores may have barely managed to produce enough offspring to maintain their numbers, nothing would have been gained by yearlong sexual activity, for ungulates and other large animals with a comparably long gestation period and seasonally limited mating produce yearly births, and woman could hardly do more than this.

Anthropologists have viewed the almost continuous sexual receptivity of the adult human female as a means for holding primitive groups together and preventing the males from fighting for women. But if, as in other animals, the brief seasonal sexual activity of the males coincided with that of the females, all might live in amity through the long months of sexual quiescence, as many of these animals do. Moreover, as has been all too evident throughout recorded history, continuous female receptivity does not prevent adultery, which in earlier times might be followed by the death of the delinquents but in this laxer age leads only to ugly squabbles and divorces. Except to the hedonist, man's overwrought sexuality, the source of so many evils, must be regarded as one of the misfortunes of his evolution.

In this matter, nature has been kinder to other animals. In mild climates where migration is unnecessary, birds of many species live in inseparable pairs during the long season when their reproductive organs have so profoundly regressed that they are practically sexless. Bands of primates, from lemurs to great gorillas, are held together by mutual liking rather than the rare upsurge of sexual activity. These birds and mammals seem to approach closer to the human ideal of Platonic love than oversexed man often succeeds in doing.

Morality

Sex, the vicious one of the pair of winged steeds that, in Plato's metaphor, draw the chariot of the soul, would cause even more disasters if it were not, in some measure, held in check by moral reins. Morality, at least up to a certain level, offers no difficulty to evolutionary theory. It is deeply rooted in the Universe, and in its most rudimentary form it seems to antedate life. The solar system itself, with its great bodies all circulating around each other in perfect equilibrium, each in its own orbit, influencing yet never harming each other, provides a perfect schema of a moral society. At the other extreme of size, the atoms or molecules in a crystal, lined up in orderly rows and layers, each contributing its share to the wholeness of a formation that is both beautiful and enduring, provide a paradigm of a moral community, although perhaps one that is overregimented.

From this moralness of the nonliving world we ascend to the protomorality of animals. Whether or not imposed by something analogous to conscience and a sense of duty, the care that many birds, mammals, and even fishes and insects lavish upon their young is often such as the most exacting moralist would approve. In their social relations, too, animals often exhibit behavior that smacks of morality. Many continuously mated birds are models of conjugal constancy.

Mammals and birds of numerous kinds hurry to assist companions or young in trouble, often at great risk to themselves. After their territorial boundaries have been settled, birds tend to respect their neighbors' domains, and when trespassing they behave in a way that suggests a sense of guilt; the trespasser is easily chased away by the very same individual whom, on its own side of the territorial boundary, it can in turn put to flight. Evidently something more subtle than brute force or fighting ability enters into these relationships.

The behavioral patterns of men in a tribal society are considerably more complex than those of birds or quadrupeds, even in the most advanced of their societies. A non-human observer obliged to study a primitive human society as objectively as we must watch animals, without any possibility of communicating by speech, might detect in it merely a further elaboration of the protomorality that we find in birds. But we who can penetrate more intimately into the psychic life of primitive men are sure that their behavior is influenced by feelings whose presence in non-human animals we may sometimes suspect but cannot demonstrate. These feelings of obligation, guilt, and shame are the voice of the collectivity speaking in the intimate depths of the individual; they are sometimes ascribed to a superego, moderating or vetoing the more selfish impulses of the individual's ego. So strong is tribal man's dependence upon the opinion of his neighbors that, if his transgressions become public, guilt or shame may impel him to destroy himself.

Man has been so successful in the struggle to survive and multiply because he has learned to cooperate with his fellows, first in such basic endeavors as procuring food, erecting shelters, and resisting enemies, later in the much more elaborate enterprises of advanced societies. Social life is hardly possible without rules for regulating the behavior of individuals, restraining their selfish or aggressive impulses; rules of this sort are the substance of primitive and even civilized morality. Since such morality has been indispensable to man's success as a social animal competing with other animals, it was promoted by the usual agents of evolution—mutation and selection—and followed naturally upon the development of a large brain and of language.

More difficult to account for is the morality that rises above that of one's tribe, social group, or inherited religion. From time to time, a man is born who feels the traditional morality of his neighbors woefully inadequate, oppressive to his sense of righteousness and justice, stifling in the narrowness of its concern. He longs to break down its tight barriers and admit into the community to which moral considerations apply not only his tribe but neighboring tribes, not only his race but alien races, not only mankind but the whole living world. By adopting principles and a mode of life at variance with those of his neighbors, he exposes himself to ridicule, to social ostracism, often to more active persecution. In striving to cultivate more harmonious relations with those farther removed from himself in origin or space, he may estrange himself from those nearest him. This open morality is devoid of any immediate survival value; at most it might anticipate a remote situation in which all men, or all life, would survive more securely. How can the harsh, gambling methods of evolution give birth to such all-embracing goodwill?

Man's Expansive Spirit

Other aspects of the human spirit exhibit the same expansiveness. The prototype of compassion is the mother bird's or mammal's tender care of her helpless offspring, her sheltering breast, her prompt responses to its need for food or warmth, her protective zeal. This same solicitous mother may be quite capable of snatching away the callow young of some other parent to feed her own family or herself. From this narrow source, compassion has expanded until, in the nobler sort of man, it embraces all that can feel and suffer, so that he will not avoidably bruise the meanest worm.

Love is allied to compassion and is of similar origin. Although we commonly think of it as the affective bond between a youth and maiden or a man and woman, actual or potential sexual partners—and so it has been most frequently celebrated by poets—this is not its original context. In its more primitive manifestations in the animal kingdom, sexual attraction is a biochemical or physiological phenomenon, devoid of tender affection and seldom persisting after it has accomplished its primary function of fertilizing the female's waiting ova. Love as a spiritual affection appears to have originated in the relation of mother and child and to have passed thence to include the father when he was closely associated with the mother in the child's care, as happens very frequently among birds and less often in mammals.

Spreading from its familial source, love may embrace things the most diverse. The little child loves, at times with passionate intensity, his pets or stuffed effigies of animals. Some of us come to regard the natural world, its trees and flowers, birds and furry creatures, hills and streams, with a feeling scarcely less intense than our love for members of our family. We may even profess to love mankind, but to do so it seems necessary to idealize humanity, to view it as it could or should be rather than as it is; in its actual state, humanity is more apt to excite pity than love.

To become familiar with its environment, with its sources of food, its perils, and its safe retreats, is a necessity for any animal more mobile than a barnacle or an oyster. Curiosity widens knowledge of the environment and often leads to the discovery and use of new sources of food, to safer shelters, or to the detection and avoidance of enemies. By no means confined to man, it grows with more acute sensory equipment and increasing mental activity. Up to a certain point, curiosity improves an animal's chances to survive and is promoted by natural selection. But man's drive to explore, his thirst for knowledge, is not limited by his practical needs. He wishes to know about distant lands, the farthest stars, the most remote past, the deeds of men long since turned to dust. He is so inquisitive about the future that he frequently resorts to the most absurd means to disclose it. He pries into the hidden depths of matter, examines the structure of living things, studies the habits of manifold creatures. Although his quest for knowledge often yields information that increases his material prosperity and helps him control his environment and spread over the world, a large share of his studies is prompted by no such utilitarian motives. Man's curiosity is expansive, like his morality, his compassion, and his love. He seeks knowledge because he loves it.

How shall we explain this expansiveness of the human spirit, which, when fully awakened, reaches out toward the infinite, yearns for union with a transcendent perfection that it calls God, loves all things good and beautiful however remote they may be, and feels compassion for the humblest suffering creature? Can evolution by random variation and the destruction of misfits account for spiritual qualities that, far from promoting survival and reproduction, all too frequently cause reduced efficiency in the struggle to survive and reproduce? Are not these qualities a revelation of something older than the evolutionary process and organic life itself, something that stirred in the primal Universe and, seeking expression but lacking guidance, had no recourse but blindly to grope its way forward by the harsh, crude, trial-and-error methods of organic evolution? Certainly an omnipotent deity, able to create precisely in accordance with his benevolent will, would not have chosen such a slow, indirect method; but a germ of divinity, lurking at the heart of nature like a seed deeply buried in the soil, might have striven to unfold itself in this blundering manner because, until it had fashioned adequate organs for itself, it could do no better.

One such organ, still imperfect, is the human mind, whose embodied activity depends upon an organic brain that had to be developed by the same process which evolved

all the other organs and functions of living things, a process
which at long last awakes to

> a sense sublime
> Of something far more deeply interfused,
> Whose dwelling is the light of setting suns,
> And the round ocean and the living air,
> And the blue sky, and in the mind of man;
> A motion and a spirit, that impels
> All thinking things, all objects of all thought,
> And rolls through all things.

These lines from Wordsworth's "Tintern Abbey" tell us as
much about the Something that set the world process in
motion and gropes its way through the evolutionary process
as we know. Neither science nor philosophy nor theology
can at present tell us more, without becoming dogmatic.

Man Not Evolution's Single Goal

Although man, at his best, might be considered the highest product of evolution on this planet, in the sense that he
seems most adequately to express and to be most fully
aware of the Something that the poet felt so keenly in one
of his more inspired moments, we must avoid the common
error of assuming that we are the unique end of the world
process, the single goal toward which all its striving has
been directed. If this process is Being's effort to enrich itself
by the full realization of the values it potentially contains,
then to suppose that man is its only goal is tantamount to
assuming that he is capable of experiencing all the values
that life can bring forth and that without him terrestrial
life would be valueless. A little reflection should dispel this
absurdly anthropocentric assumption. No single man can
appreciate all the true values available to humankind: the
sedentary person can hardly imagine the joys of the mountaineer, and the latter may doubt that anyone could enjoy
solving the abstruse problems of pure mathematics. How,
then, can we hope to assess all the satisfactions that make
life precious to all the living things on Earth?

Man is but one among many achievements of evolution, none of which can be considered final so long as this process continues. What would this planet be like without its covering of green plants? That they have indispensable instrumental value in supporting animals is obvious; but we must not too hastily assume, on purely negative evidence, that they derive no satisfaction from growing and absorbing sunshine. The aerial life of birds is a revelation of life's creative power no less wonderful than man's brain and technological skill; it would be most surprising if creatures so richly endowed with flight, beautiful plumage, song, and sociality led a joyless existence. We have as yet only intriguing glimpses into the advanced psychic life of dolphins and other cetaceans. Who shall assess the satisfactions available to the teeming tribes of insects or the multitudinous creatures lower on the evolutionary scale? Life is a tree of many branches, many of which are necessary for its symmetry and continued health; one branch cannot dominate the tree, directing all its growth into itself, without unbalancing it and finally toppling it over.

Although man has accumulated the most conspicuous prizes in the long evolutionary game, it would be a mistake to conclude that all the winnings are his. Many other players, from insects to birds and mammals, have won very considerable stakes. Indeed, any contestant that remains in the game to this day, down to the simplest worm, must have won something, if only a consolation prize. The losers are those innumerable species that have become extinct through the immense geologic ages, without giving rise to newer species. And even they may once have held winning cards and found some enjoyment in the game while they managed to remain in it. Moreover, we cannot be sure that the players who are ahead today will not lose everything tomorrow, for the game will probably continue as long as life flourishes on this planet.

Like any uncontrolled movement, evolution has frequently gone astray, yielding disvalues instead of values, suffering as well as delight. Some of its products, especially noxious parasites, evidently cause an amount of suffering that far outweighs any satisfactions they are likely to experience in their degenerate existence. In our less sanguine moments, we suspect that man must be included among those creatures that bring more disvalue than value into the realm of life. Despite his splendid endowments and capacity for joyous living, he has generated a vast amount of misery among his own kind by his inability to control his disruptive passions and his inept social arrangements, and to this must be added all the havoc that his greed and aggressiveness cause in the rest of the living world.

Summation

To summarize this long chapter, of all the animals on this planet, man has probably the richest hereditary endowment, in his body capable of varied activities, his fine complement of sensory organs, and his superior mind, along with a potential life span far exceeding that of many mammals much bigger than himself. These advantages greatly outweigh certain defects due to his retention of vestigial organs that are no longer useful and to structural weaknesses that result from his recent (in an evolutionary sense) assumption of an upright posture. Man owes these excellent innate qualities not to any intelligent planning or other laudable effort of his own but to a long run of luck in the hazardous evolutionary game.

The first of these favorable turns of fortune's wheel occurred ages ago, when certain small, primitive mammals adopted an arboreal life, climbing by seizing branches in their digits rather than grappling them with sharp claws, thereby giving rise to the primate stock to which man belongs. The second outstanding piece of luck in man's evolutionary history was the return of his ancestors to the ground, after many millions of years of arboreal existence, with the capacity to walk erect, thereby freeing their forelimbs for varied uses. Man's large brain and his speech evolved in closest correlation with his versatile hands, which needed a mind to guide them, along with social cooperation for which language was indispensable, in order to realize their full capacity for contributing to his welfare.

Man did not emerge from the long evolutionary struggle without acquiring certain psychic defects that have caused much more suffering than such corporeal imperfections as a vermiform appendix subject to infection and wisdom teeth liable to become impacted. Chief among these are his often violent passions, his aggressiveness, greed, cruelty, hatred, and lust. His too-powerful sexual impulses and yearlong reproductive activity have been a major cause of human misery and are responsible for population problems that threaten to wreck the planet if not speedily solved. These psychic defects and physiological excesses might be attributed to the trauma of evolution.

On the other hand, man has certain psychic qualities difficult to account for by evolutionary theory, since they make no immediate contribution to his survival. The expansiveness of the fully awakened human spirit, as revealed by love and compassion that extend far beyond our own kind, thirst for knowledge of distant things, yearning for the infinite and an absolute perfection called God, is evidently an expression of something that is far older and more deeply rooted in the Universe than the processes of organic evolution and is the driving force behind them.

Despite his advantages and achievements, man must not regard himself as the unique end of the world process on this planet, since he is not capable of realizing all the values that it may generate here. After thousands of millions of years of activity with blundering methods, evolution has achieved magnificent results in many directions. Life is a tree of many branches, none of which can become overgrown at the expense of the rest without destroying its symmetry and balance.

9. The Ascent of Consciousness

 A world without beings that feel and enjoy would appear to lack significance. No matter how fair its skies, how balmy its air, how green its landscapes, without sentient inhabitants it would seem as desolate as the Sun-scorched, lifeless spheres of Mercury and the Moon. All the value in the Universe—everything that, as far as we can tell, gives it meaning and worth—depends upon its psychic aspect, which, unfortunately, is far less evident and accessible to investigation than its ponderous physical aspect.

To form a fair estimate of the Universe, or of our own planet, we would need to know far more than we do about the distribution of consciousness and of its more primitive state, bare sentience or feeling devoid of thought. Each of us is immediately aware of his own conscious life, and we infer, although we cannot demonstrate this by scientific procedures or prove it by irrefragable logic, that other people have feelings and thoughts somewhat similar to ours. We spontaneously attribute corresponding emotions, if not thoughts, to the animals most akin to ourselves, especially warm-blooded mammals and birds and, above all, those most intimate with us and responsive to our kindness. Again, proof that we are right is lacking.

The more dissimilar to ourselves a creature is, the less confident we are about its psychic life. We are doubtful about the feelings of cold-blooded reptiles, amphibians, and fishes, even more doubtful about those of insects and other invertebrates. Until recently, only poets and mystics attributed feeling to the vegetable kingdom; but certain recent observations of their electrical responses to injuries and even to threats of harm, although falling short of convincing proof that plants have a psychic life, should at least make us keep our minds open to this possibility.

Many thinkers have attributed feeling even to lifeless matter. If atoms were wholly devoid of feeling, it would be difficult to conceive how feeling arises in us who are made of them. Hard denial of the possibility that atoms feel comes chiefly from those who, from the admission of their sentience, jump to the conclusion that they must therefore have emotions similar to our own—a notion that Lucretius ridiculed in a famous passage in the second book of his poem *On the Nature of Things*. Just as a single atom's weight is too small to be noticed by us, so its feeling might be so slight that, if added to our massive consciousness, it would pass undetected; yet even such minimal feeling might have important consequences for cosmic evolution. Since all but an infinitesimal fraction of matter persists in the inorganic state, a Universe in which lifeless matter finds satisfaction in its existence would have greater significance, and be more pleasing to contemplate, than one composed of wholly insentient atoms.

However it might be at the lower levels of integration, at the higher levels we can recognize three ascending degrees of consciousness and of the values that it enjoys. The first is contentment or joy springing directly from one's own existence or activities. Conceivably, plants feel this, at least dimly, as they bask in the life-giving sunshine; but, lacking a brain or a central sensorium and an integrating nervous system, they could hardly enjoy unified consciousness. Each cell, or at most each leaf, might have its own feeling. Likewise, it is conceivable that the cells or organs of our bodies are not devoid of feelings and that these feelings contribute to our gladness when we are in fullest health and help depress us when vitality is impaired. Sometimes, when on a sparkling day I walk in woods or fields, I wonder whether my joy does not owe something directly to the pleasant feelings of the life around me.

These speculations aside, it is hard to doubt that the more advanced animals, especially birds and mammals, feel contentment when healthy, well fed, and in a congenial environment and that their pleasure can be increased by such activity as eating, singing, or playfully romping. Horses, for example, gallop about the pasture for no apparent reason except the joy of exercising their strength in exhilarating motion. Even when well fed, they whinny for special foods in a manner that clearly reveals pleasant anticipation. Wide-winged birds soar on ascending columns of air in a way that suggests pure enjoyment. And we cannot doubt that children enjoy their play, their rides, their sweets, and much else. Simple contentment with our own existence and activities fills many of our waking hours, when all is well with us.

Consciousness rises to another, higher level when it is enhanced by contemplation of the things around us. Above all, beauty attracts us to the surrounding world and increases our satisfaction with existence. We seem to be made to contemplate beauty; in the natural world we see it everywhere, from the creatures that, through a microscope, we watch swimming in a drop of water to the most stately trees. Grandeur and sublimity can stir us more deeply, whether we gaze upon a snow-crowned peak rising above tropical forests, peer into the colorful depths of a profound canyon, or gaze in wonder at the starry sky. To watch children happily at play, people engaged without strain in satisfying tasks, birds songfully building their nests or feeding their young, bees gathering nectar or pollen, dolphins racing ahead of an advancing ship, and many another natural sight is a source of enjoyment that draws consciousness beyond its own body.

One great advantage that we have over animals is that we enjoy watching them as they apparently rarely or never enjoy watching us. They commonly enrich our lives as we seldom enhance theirs. Merely to feel the presence of a loved one is a high value, as is contemplation of a truly good or noble person, at whatever distance in space or time he lives or lived. Perhaps it would not be too farfetched to add here contemplation of ideas, concepts, mathematical or scientific truths, and similar mental contents. Although they exist in the mind, they draw it beyond the body. I suspect that it was this sort of contemplation, rather than admiration of beautiful objects and other corporeal things, that Aristotle meant when he declared contemplation to be the highest good.

Finally, the highest level that consciousness has yet attained on our planet appears to be sympathetic participation in the psychic state of other beings. When we watch active animals, we may delight in their graceful forms and motions—which is consciousness of the second sort—or we may go beyond this point and try to feel as they do. Similarly, when we watch children contentedly playing, we may join them in spirit by trying to resurrect the delights of our own childhood, which may be almost as difficult as attempting to feel as animals do.

Sympathetic participation extends to pains and sufferings as well as to joys, and perhaps in this sphere the induced state is a truer copy of the original, for pain appears to be more primitive or undifferentiated than pleasure; and suffering, more than joy, binds all flesh together. This may be because we often envy the joys of others but can only commiserate with their pains. Although to multiply suffering by sympathetic participation diminishes the value of existence, the capacity to sympathize with other beings, whether in joy or in sorrow, reveals a very high development of consciousness; and, as a source of active compassion, the tendency to suffer simply because others suffer mitigates life's harshness. This third stage of consciousness is still poorly developed, even in the most sensitive people. Our sympathetic representation of the psychic states even of those closest to us can be far from accurate. When we try to extend this process to nonhuman animals and beyond, we are on extremely hazardous ground. Nevertheless, I am convinced that a deeper understanding of the Universe awaits the evolutionary extension and intensification of this third stage of consciousness.

These three levels of consciousness are not mutually exclusive but additive and reciprocally reinforcing. If we did not enjoy our own existence and activities, we could hardly imagine that other beings do so. If we were not strongly attracted to them by their beauty, grace, or interesting habits, we would have less incentive to speculate about their psychic life. To suspect or to have faith that creatures the most diverse enjoy and suffer, even if not as acutely as we do, binds us closer to them in sympathy. Each of these kinds of consciousness is enhanced by grateful appreciation of the privilege of living, of the pleasure and absorbing interest of contemplating the beauty and grandeur of the cosmos, of the sympathy that overleaps the biologically necessary insulation of organisms and draws us into closer union.

Consider a creature that lives happily, keenly enjoying nature's bounty, neither harming anything nor feeling any responsibility for preserving anything, and without gratitude for its existence and all that enriches it—a creature that lives like an innocent, joyous child or, perhaps, some innocuous animal with keen senses and a great capacity for pleasure but little thought. Would not the presence of such a creature augment the value and significance of the Universe? Add to this creature gratitude for all that it enjoys. Is not the value of the Universe still further increased? Add again a sense of responsibility for preserving and perhaps improving whatever gives it joy, and the value of the Universe rises by another degree.

In these conscious states, the agelong striving to raise bare existence to meaningful existence reaches partial fulfillment. Even if nothing ever paid the least attention to anything else, a world in which each creature felt a modicum of enjoyment or satisfaction in its own existence would be vastly superior to one composed wholly of dead, feelingless matter. But appreciative contemplation immensely enriches the world. Imagine a sensitive naturalist reaching a remote mountaintop where human foot has never trodden. Around him bloom flowers, perhaps kinds unknown to science, "wasting their sweetness on the desert air." Lovely birds of rare or unnamed kinds flit and sing amid the shrubbery. Even if flowering plants and birds enjoy their existence, the naturalist's appreciative delight in them increases their value and imparts to the mountaintop a significance that it hitherto lacked. If he writes feelingly about his experience or publishes paintings or photographs of what he sees, the formerly unknown mountain enters the ongoing stream of human consciousness. Although it is doubtful whether birds or flowering plants feel better or happier for being at last admired and loved, the complex of man plus plant and bird acquires higher value, and the world is enriched thereby.

What the sensitive naturalist was to the mountain, so are appreciative human minds to the cosmos or to our small part of it. They lift creation to higher levels of significance and immensely increase its total value. Without minds to contemplate, appreciate, and strive to understand the accomplishments of harmonization, the cosmos would remain incomplete, its splendid potentialities still unrealized. It would be like a seed that never grew into a flowering plant, a butterfly that never escaped its cocoon to spread satiny wings to the sunshine. We can only surmise to what supernal heights creation may have risen on planets that orbit around distant stars, or what it will eventually accomplish on our own. At present, minds that gratefully appreciate, love, and care intensely about their planet and its life must be regarded as harmonization's finest achievement, because they so greatly increase the worth of everything lovely and good that it has produced.

10. Beauty, the Basic Value

While filling Earth with life, evolution covered it with beauty. Not the least of its precious gifts to us is our capacity to perceive beauty, not only in the living world but also in the inorganic world that supports it and the heavens that surround it, and to be refreshed and uplifted by this vision. Nothing so swiftly excites our admiration and appreciation as beauty. As we travel through Earth's wilder regions, far from the crowded centers of man, we detect beauty and grandeur in a hundred situations where we could survive only precariously or not at all, in a thousand objects of no practical use to us, even in some that might injure or destroy us if closely approached. In the frozen polar regions, in arid deserts, in steamy tropical forests, where we can exist only with difficulty and at the price of enduring hardships, we meet beauty on every side: in the gleaming snow and spectacular atmospheric phenomena of the Arctic; in the colorful rocks and glorious sunsets of the desert; in the forms and colors of plants, birds, insects, and other animals of the tropical selva.

Through powerful telescopes we gaze into the depths of space and discover beauties that the starry heavens withhold from our unaided vision. Through a microscope we examine a drop of pond water or the tissues of a plant, and here, too, marvelously beautiful patterns greet our delighted gaze. Even the bright colors whereby certain insects and other animals warn us that they will hurt or destroy us if we touch them may attract us by their beauty. The aesthetic congeniality of Earth contrasts with its frequent indifference or hostility to our practical interests. Our aesthetic adaptation to the wider world is superior to our biological or social adaptation; it is the most perfect that we have achieved. We seem to have been created to enjoy beauty.

I sat on a stone in the bed of a mountain torrent, in the dry season when the stream had dwindled to a trickle between small pools. Before me the gray boulders, of all sizes and shapes, had been heaped helter-skelter by the floodwaters of many a rainy season. Trees along the banks leaned over the watercourse at diverse angles. Here was neither order nor symmetry nor any color brighter than the dun rocks and the green of the foliage that shaded them. How could a mind that delights in symmetry and balance, as in well-proportioned buildings, flowers, or graceful animals, be other than repelled by such huge disarray? How could any vision that enjoys color be held by the somber shades of these rocks? Nevertheless, I found this wild scene so attractive that I sat a long while contemplating it, tracing with my eyes the shapes of the boulders, admiring the random way that they were piled together, and catching glints of sunshine reflected from the shallow pools, where tiny fishes swam. Finally I arose, to continue my laborious progress over the tumbled rocks, deeply impressed by the breadth and versatility of our aesthetic adaptation to the natural world, which prepares us to respond with delight to the most rugged scenes no less than to the fragile loveliness of flowers.

Thus nature, which frequently repels us by its harshness, wins us back with its beauty. Our planet owes its loveliness, in the large and in detail, above all to its mantle of green plants. Not only are we indebted to them for a very large share of Earth's beauty, we owe to them, above all to trees, much that contributes to our enjoyment of beauty, even our ability to create it with our hands. In all probability, we have color vision, which many other mammals lack, because our remote ancestors were diurnal, frugivorous forest dwellers who, like birds and certain rodents, helped scatter the seeds of fruits which made themselves conspicuous by colors that contrasted with the prevailing verdure, as well as by their aroma. Moreover, to gauge the distance of their often prodigious leaps from bough to bough, these arboreal ancestors, like other primates, needed highly efficient stereoscopic vision, which is best achieved by eyes directed straight forward rather than more or less sideward, as in many vertebrates.

To be able to focus our vision sharply upon a symmetrical, three-dimensional, colored object helps us greatly to appreciate its beauty, but something else appears to be needed: an alert, intelligent mind, whose interests range beyond the basic needs of food, safety, and reproduction. For such minds we are, as was explained in chapter 8, at least indirectly indebted to the trees in which our ancestors developed the grasping hands that also enable us to create beauty. Thus it appears that our aesthetic responsiveness, like so many other precious things, is an outcome of constructive interactions between animals and plants.

But why should we be sensitive to beauty at all? Our large brains and stereoscopic color vision present no difficulty to the evolutionist, because he can show that they promoted survival at various stages of human evolution. However, these are only the apparatus that enables us to perceive beauty; they are not aesthetic sensibility itself, for which some further explanation is needed. Can this be more than an incidental development of a psychophysical organization that evolved primarily for such practical ends as finding food and shelter, escaping enemies, and the like? Aesthetic sensibility has its negative as well as its positive aspect; the counterpart of the capacity to be agreeably excited by sensations that do not directly promote survival is susceptibility to depression by sensations that are neither harmful in themselves nor signs of something injurious— by ugliness itself.

Given such sensibility, however it arose, natural selection would, I believe, promote its positive response to certain objects and situations to which the animal was constantly exposed. If the azure of the sky and the verdure of the terrain were as depressing to an animal as certain drab colors can be to us, its vital processes and its will to live might be adversely affected, so that in the struggle for existence it would be less successful than some related animal who, instead of being depressed, was pleasantly excited by these widespread colors. It seems obvious that, when animals choose their sexual partners by visual characters, an individual will be more likely to mate and reproduce if it is attracted rather than repelled by the appearance of potential partners. The reality of the aesthetic response is assumed by the theory that Darwin developed to account for the more striking adornments of animals by sexual selection. Thus, widely accepted biological principles seem adequate to explain why we, and doubtless other animals with well-developed vision, find beauty in a blue sky dappled with white clouds, the green and flowery Earth, and members of the same species, especially those of the opposite sex in their prime.

I doubt, however, that the foregoing considerations can adequately account for the whole range of our aesthetic responses. They seem especially inadequate in the case of music, which poses one of the most difficult problems to any theory of aesthetics. Why should a complex train of atmospheric vibrations, conveying no welcome information and contributing nothing material to our welfare, delight us so powerfully as music can do? I believe that we can account for this only by the fundamental principle that harmony is the source of all joy and happiness, not only in ourselves but doubtless also in every conscious being.

Harmony arises whenever a number of diverse parts form a coherent pattern; the harmonious object is simultaneously many and one. In the case of music, the component parts are notes of diverse pitches and intensities, joined in a coherent pattern that we call a melody. In a painting or a landscape, the components are colored forms, which must be combined with a certain coherence and balance to appear beautiful to us. Hardly anything is more essential to our felicity than health, which we enjoy when all the organs and functions of our bodies work together to make one harmonious whole. But, above all, happiness requires the harmonious integration of our spiritual life in all its aspects, as well as the regulation of our activities in conformity with our ideals, aspirations, and principles of conduct. One might adduce many other examples of the general rule that we experience pleasure or happiness when the objects or situations that affect us form harmonious patterns; and, when these patterns are perceived by means of our external senses, we call the resulting pleasure aesthetic and say that we enjoy beauty.

Since to form a harmonious pattern is the condition of beauty, and organisms can hardly survive unless they are harmoniously integrated, it is not surprising that beauty is widespread in the living world. Yet this alone will hardly explain why we discern beauty in so many things that contribute nothing material to our welfare and may even be hostile to us. May not the reason be that, simply to enjoy beauty, we demand nothing of the beautiful thing, except that it be itself? Being itself, it is beautiful; and being beautiful, it cheers us and fortifies our will to live.

In most other relations with objects that somehow reward us, we make greater demands on them. To serve us, they must deviate from their usual course, make an effort, or relinquish something, perhaps their lives, as often happens when we use them for food, clothing, or the construction of our houses. But everything, especially every living thing, exists primarily for its own sake rather than for another's. If an animal, it may resist, with fangs, horns, or other weapons, our attempt to exploit it; if a plant, it may cover itself with thorns or stinging hairs or impregnate its tissues with distasteful or poisonous substances to make itself inedible. But, when anything can enrich our lives simply by being itself and following its own course, it has no reason to withhold this boon. When we require of the beautiful object only that it be itself, we find beauty everywhere.

This fact should regulate our treatment of beautiful things. When young and foolish, we wish to seize and retain whatever delights us, so that, as we fondly imagine, we may enjoy it forever. Many, unfortunately, never outgrow this childish attitude. Such thoughtless possessiveness can make beauty a source of envy and strife among men and a curse to outstandingly beautiful animals and plants. Helen's beauty led to Troy's destruction, and the loveliness of certain birds and wildflowers has jeopardized their survival. With growing insight, we may realize that the proper treatment of something that rewards us simply by being itself is to permit it to remain itself, interfering with it as little as possible. Since it is wrong to claim for oneself alone what is freely offered to all, we should protect the beautiful object without seeking exclusive possession of it. In a society not tainted to the roots by the poison of commercialism, things cherished for beauty alone would not become objects of speculative trade. Artists would create them for the joy of creation and place them where they would be most appreciated.

Truth, beauty, and goodness have, since ancient times, been named as the three great categories of the higher values. Interpreted broadly, these overlap—for truth and beauty are certainly forms of goodness in the widest sense; the good is often said to be beautiful; and poets have proclaimed that beauty and truth are one. But the boundaries between the three become sharp when we restrict truth to the realm of the intellect, beauty to that of sensuous impressions, and goodness to the will or to our moral determinations. So defined, beauty is the most widespread, primitive, and fundamental of the three great categories of values. It is by far the most abundant and easiest to enjoy; in the natural world, it flows in to us from every side. Although appreciation of beauty is, like other innate capacities, heightened by cultivation, even in its native form, as in the child and the savage, it is a source of delight. Goodness and truth rarely come to us spontaneously, as beauty so often does. As Pittacus remarked long ago, it is hard to be good. Truth, especially about the fundamental problems that most concern us, is even harder to attain than goodness. The road to truth is so steep and stony that few travel far along it, and even the stoutest and most resolute seekers falter and perish before reaching the summit.

The reason why beauty is the most widespread and basic of the higher values seems obvious. If the creative process had not first enriched our planet with beauty, it might never have led on to truth and moral goodness, because in a world devoid of beauty we might have little incentive to discover truth or to become good. Many of us are stimulated to learn about Earth, to disclose the secrets of nature, because it attracts us by its beauty—the grandeur of the starry heavens, the loveliness of plants, the graceful forms and brilliant colors of animals of many kinds. If these things repulsed us by their ugliness, we might not care to know about them. It might be contended that, even if our world were wholly devoid of beauty, we would still try to understand it because knowledge of nature helps us survive. But would we wish to survive on a planet devoid of beauty? If, wherever we turned our eyes, they fell only upon drab, unlovely things, if we heard only raucous sounds, if every scent that greeted our nostrils were a stench, and we had no hope of escaping to fairer skies, would we desire to live another day?

Just as, on a planet without beauty, we might have little incentive to acquire knowledge or truth, so might we lack strong motives to be good. Often the beauty or attractiveness of the people or other creatures that we meet incites us to cultivate harmonious relations with them, which is a large part of what we mean by moral goodness. Moreover, we are good because we love life, and the long experience of mankind has taught us that life flourishes most, and yields the most satisfying experiences, when the fundamental moral rules are most carefully obeyed. But without beauty, life would lose its tone; it would have little value for us; and not caring whether we lived or died, we would hardly make the effort to become good.

Beauty, then, was needed to prepare us for the arduous quest for truth and the strenuous effort to become good. Animals who apparently never try to understand the causes of things, who seem to have no reflective morality but only an innate protomorality, are nevertheless responsive to beauty, as is evident from the elaborate nuptial adornments of many of them—adornments that could hardly have evolved unless individuals of the opposite sex were favorably impressed by them. Savages who live in filth and squalor, who give free rein to the most violent passions, whose morality extends no farther than their own tribe, who are rarely moved by sympathy and pity, whose science and philosophy are rudimentary, who are perhaps headhunters and cannibals—even such rude people not only appreciate beauty but often try, not unsuccessfully, to create it, adorning their utensils or weapons in a way that may seem utterly incongruous with the surroundings in which they take form. Only at higher stages of culture, when people become more sensitive and thoughtful, do they feel oppressed by their deficiency in knowledge and goodness. Then they are often appalled by the contrast between the splendor of their planet and the strife that pervades it, daily destroying countless lovely creatures, filling with pain and anguish a world that, one might suppose, was created chiefly for the enjoyment of its beauty.

Our aesthetic adaptation to the world in which we live is, as we have seen, much superior to our biological and social adaptation. Beauty is far more widespread than its sister values of truth and goodness. Appreciation of beauty and even some ability to create it arise in children and savages, whose intellectual and moral development is still rudimentary. These facts suggest the conclusion that, in the course of our evolution, the aesthetic side of our complex nature has raced ahead of other equally important aspects of ourselves, such as the cognitive and the moral; that we have become, above all, instruments or organs for the perception of beauty. Could it be that just this capacity is our most significant contribution to the Whole to which we belong?

The world process is most profitably regarded as Being's striving for self-realization, a titanic effort to understand itself and to bring forth and make actual all the beauty, all the goodness, all the joy, all the values of every kind that were latent in its primitive state. Just as an animal cannot see and hear without functioning visual and auditory organs, so the Universe could not actualize the values that were latent in it without creating, from its own substance, organs for knowing, appreciating, enjoying. We are such organs—still very imperfect ones, to be sure, but evidently the most adequate that have arisen on this particular planet. When we enjoy beauty, as when we know and understand, we are the Universe appreciating and understanding itself by means of organs which, by means of aeonian striving, it has created for this purpose. This appears to be our proper function, our greatest contribution to the Whole to which we belong. For what would have been gained by the evolution of countless harmonious forms, on Earth as in the heavens, if there were no creature able to appreciate them? Certainly a world that contains much potential beauty would lack completeness without organs capable of actualizing it—for beauty unperceived, as in the flower that blooms unseen in the wilderness, is not beauty at all but only its potentiality.

To be sure, the flower has a certain symmetrical form, which all but the most stubborn skeptic would admit exists extended in space, even in the absence of any observer. It reflects rays of light that consist of pulses or vibrations that differ in frequency and wavelength. It diffuses into the air certain molecules of complex structure. This is all that physical analysis reveals of the flower's claim to loveliness and fragrance. You admire the blossom. Streaming through the pupils of your eyes, a minute fraction of the luminous rays that it reflects is focused upon the corresponding retina in a pattern that reproduces, in miniature, that of the flower itself, much as light is focused by the lens of a camera upon the sensitive film. Transmitted almost instantaneously by the optic nerves to your brain, the excitations that originate in the rods and cones of two separate retinas are, by a marvelous transformation that we are far from understanding, united in a single, delicately colored mental image. By another transformation that seems less complex, molecules that diffuse from the flower to your nostrils are perceived as a delicious fragrance. But, unless you are attentive and in the proper mood, you may fail to be pleasantly excited by the form, color, and odor that you perceive. An aesthetic response depends upon subtle psychic factors.

From the foregoing analysis, it appears that the flower, in itself an organic structure with certain physical and chemical properties and perhaps a measure of sentience, becomes beautiful at the moment when it is perceived by a sensitive onlooker. Its beauty does not reside in the flower alone, nor yet in the observer alone, but is created by their interaction. Every perception of beauty is a fresh creation, born of the fertile union of an appropriate object and a properly equipped spectator. To this creative synthesis, the latter appears to make the larger contribution. A flowering plant or the most elaborate work of art is a structure far simpler than the manifold of eyes, nerves, and brain by means of which we perceive and respond to it. It is hardly an exaggeration to say that we find nature beautiful because we, who are parts of nature, clothe it with beauty.

Thus beauty, like other values, is the product of the interaction of a value generator, such as a flower, and a value enjoyer, or organ of appreciation, such as a sensitive human spectator. Were any of us an adequate organ for the appreciation of the values that our planet offers, in the sense that he could realize them all, fully and permanently, his whole effort should be directed to the exercise of this, his proper function. For him, morality might consist in nothing more than responding to the beauty of the world and perhaps also understanding it and, of course, in being careful not to diminish the values that he found here. But, since the best of us is a very inadequate organ and, moreover, one that will last only a short while, we must share our function with many more or less similar organs, and we must provide other organs to replace ourselves after we are worn out. Thus we might derive a whole system of ethics, of the highest order, from the premises that our proper role on the planet is to both perceive and appreciate its beauty, but none of us can fill this role so adequately that others are not needed.

When we remember that altruism does not consist only of giving valuable things or working for others but may take forms less substantial, such as words of encouragement or even a timely smile, we may recognize in beauty, wherever found, a sort of unconscious or unpremeditated altruism—for the beautiful thing, living or lifeless, enhances our lives by its presence. When we survey the great variety of things that we designate as beautiful and the strong contrasts between them, we must conclude that there is no specific quality belonging to all beautiful things but that whatever pleases us or elevates our lives we are apt to call beautiful. Beauty is our delighted awareness, directly by means of sensation, of other existing things. It is the most widespread attribute whereby one thing gives joy to another, often at a distance. Beings that can contribute no practical or material support to others may, nevertheless, enrich the consciousness of these others by beauty. Accordingly, beauty might be recognized as the primary social factor in the Universe, binding its parts together spiritually, as gravitation does physically.

The most widespread kind of beauty, the prototype of all its other varieties, is visual beauty. Light, which brings this beauty to us, is accordingly the great socializing agent of the Universe, joining even distant things in a sort of psychic union founded upon beauty, which is the spiritual cement of the cosmos. Next in frequency to visual beauty is auditory beauty, delightful sounds of all kinds. Hearing, like vision, is caused by wavelike vibrations in a medium, so that we might say that undulations are to the realm of the spirit what gravitation is to the realm of matter.

The universal law which the interpretation of beauty discloses is that the existence of every being is exalted by harmonious coexistence with other beings. Awareness of other beings that coexist harmoniously with ourselves is comforting and frequently delightful. Since morality is, above all, the effort to achieve harmonious coexistence with all that surrounds us, appreciation of beauty, without possessiveness, can be a powerful support to morality.

Of all the achievements of harmonization, none is more wonderful and admirable than the fact that some of its creations delight others quite different, who are so constituted that they can enjoy the mere presence of the former, can love and protect them, as we are charmed by birds, flowers, stately trees, and other lovely, innocent creatures. Such an outcome of evolution is so difficult to account for by currently fashionable theories that we must seek more profound explanations. This grand accomplishment of evolution suggests what its supreme achievement would be: a living world in which every creature would be pleased or comforted by every other creature that it might know, and none would shrink away, physically or spiritually, from any other. Obviously, we are still far from the realization of this ideal state.

Artistic creation is frequently regarded as one of the highest expressions of the human spirit, yet, as was earlier noticed, we find art that is far from contemptible among savages who live meanly and squalidly, with a rudimentary morality. Love of order, cleanliness, and moderation certainly appears later in human cultural development than love of beauty; hence it is the expression of a more highly developed mind. This appears to have been the opinion of

Plato, who, at the conclusion of his quest of the highest good in the dialogue *Philebus*, awarded first place to "measure, moderation, fitness, and all that is to be considered similar to these." Next, in descending order, he placed "proportion, beauty, perfection, sufficiency, and all that belongs to this class." Or might we not say that the highest art is the art of living harmoniously, with ourselves and everything that surrounds us, and that to excel in this art we need gifts of a higher order than are required to combine lines and colors or varying tones in pleasing patterns?

Often we despair of diminishing the discord to which we are exposed and making Earth as peaceful as it is beautiful. But, when we recognize the primacy of beauty among values, its profusion on Earth and in the heavens should give us hope. The first need of the world process was to generate beauty, as an incentive to cultivate with the creatures that delight us the harmonious relations that are the essence of goodness and to learn the truth about them. Now that our planet has become so beautiful, the need to increase knowledge and goodness, above all the latter, has become painfully evident. The beauty that we so abundantly behold in the natural world should fortify our faith that the process so outstandingly successful in the first stage of the creation of the higher values—the process that works in and through us—will not fail in the tasks that remain before it if we cooperate intelligently, wholeheartedly, and unitedly with it. Yet, broadly viewed, the end of all these tasks is to make the world more lovely in all its aspects, the spiritual no less than the sensuous, and to improve the quality of its organs for the appreciation of this beauty.

11. The Sources of Our Freedom

While raising life to higher levels of organization and awareness, evolution has endowed us with a generous measure of freedom. We are able to plan and regulate our lives as, it appears, no other animal can. If we would, we might, in no small degree, direct the course of man's future evolution, because we are no longer strictly controlled by the same unforeseeing processes that have determined evolution's whole past course. We are free to choose as, so far as we can tell, nothing else under the Sun is. Whether this inestimable boon or our responsiveness to beauty is the more precious of our natural endowments, I shall not undertake to decide, but they are certainly very different. The perception of beauty is immediate and indubitable. A more refined taste might alter our judgment of what is lovely; but, while it lasts, our delight in the object that we find beautiful is an unassailable fact. Freedom, on the contrary, is perhaps the least understood of our endowments, the source of great perplexity and acrimonious debates. Misconceptions about volition, or willing, have led to endless sterile arguments and prevented recognition of the true source of our freedom.

Repeatedly rejected by eminent thinkers, the doctrine of free will has been vigorously defended by other eminent thinkers and, more stubbornly, by the unphilosophical who fail to understand the subtle nature of the dispute but imagine that their human dignity and moral responsibility depend upon having free will. That our spontaneous volitions are our own and not another's is too obvious to be doubted. We are free to will as we please, even if external obstacles or our own limitations prevent the execution of our volitions. The debate about free will hinges upon the *determination* of our volitions, whether they are the necessary outcome of preceding mental and/or physical states, or whether they escape the chain of causation. A free volition is sometimes called an uncaused cause of action. It is held to be unconstrained by preceding states or events.

To understand what this means, we must look closely at causation. It is not something that we can see or feel but a concept that has grown out of our experience of nature's regularities. If B always follows A, we say that A is the cause of B; or, more precisely stated, when the pertinent conditions are exactly the same as occurred in some preceding situation, we expect the outcome to be the same. In the ultimate analysis, we cannot explain why this should be so; it is a stubborn fact of our experience, a fact upon which science, technology, the practical arts, and rational everyday behavior are all based.

Causation is indispensable for the integrity of the Universe, which without it would hardly be a Universe but a freakish play of random events. Causation might be included, together with gravitation and electromagnetic waves such as light and radio waves, among the means whereby one thing communicates with or influences another. As gravitation and electrical forces bind together contemporaneous things, so causation links together successive events. It is the bond between the past and the present, the present and the future—the line of communication down the ages. Without causation we would be utterly bewildered, because anything might follow anything else. No matter what we did, we could never be confident of the outcome.

The confidence with which chemists prepare complex pharmaceutical compounds that they sell to a trusting public, no less than the regularity of the movements of the planets and their satellites, convinces us that we live in an orderly Universe where, for the most part, causation reigns. Whether there occur occasional exceptions to the regular sequence of events has been, since ancient times, a debated question. The Stoics were strict determinists, admitting no interruptions of the causal sequence. Their rivals, the Epicureans, thought that random sideward swerves of atoms falling through the void provided a basis for free will. Modern scientists' early advances in astronomy and physics, based upon careful measurements and mathematics, encouraged the belief that we live in a world so thoroughly ruled by causation that, if we knew every dart and spin of every atom and the laws that govern their movements, we might, in theory, predict in minute detail the whole future of the Universe.

In the present century, scientific and philosophical thought, influenced largely by subatomic physics, has veered strongly in the opposite direction, toward indeterminism in the material world and its correlate in the psychic world, free will. Heisenberg's "uncertainty principle," which recognizes the difficulty of simultaneously determining both the position and the velocity of an electron orbiting an atomic nucleus, appears to challenge causation at the very foundation of the physical world. Many of the laws of physics are now held to be statistical. For example, Boyle's law, which relates the pressure of a gas at a constant temperature to its volume, depends for its accuracy upon the average force of the impacts upon the walls of the container of a vast number of atoms or molecules darting independently in different directions at varying velocities. If we could make our pressure gauge so small that only a few at a time would impinge upon it, its readings might fluctuate wildly. Conceivably, the courses of these molecules are sometimes or always indeterminate, rather than being the necessary outcomes of their velocities and interactions with their neighbors. However, I suspect that even statistical laws would become unreliable if indeterminacy frequently entered into the movements of the molecules.

The controversy between determinist and indeterminist, already well over two millennia old, cannot be settled by philosophy, theology, introspection, or scientific theory but only empirically, by precise measurements. The difficulties of such measurements are enormous, because the wide applicability of the long-established laws of physics assures us that indeterminate events are either infrequent, or of small magnitude, or both. Any measurement that appeared to support indeterminacy could be challenged by the determinist on the ground that, in nature's vast complexity, some obscure causal factor had been neglected. Contrariwise, the supporter of indeterminism might contend that measurements which appeared to support the determinist's stand were not sufficiently precise to detect small intrusions of uncaused events. After pondering this matter for many years, I am convinced that the controversy between determinists and indeterminists is unresolved and probably unresolvable but that, on the whole, we live in a determinate Universe. Accordingly, I shall not weary the reader by reviewing the endless arguments which, over the centuries, subtle minds have devised for and against indeterminacy and free will, citing such respected names as Whitehead on the side of indeterminacy and Einstein in support of determinism.

Since we cannot exclude the possibility of lapses of causation, or indeterminate events, in the physical world, where exact measurements are widely practiced, it would be presumptuous to declare that such events never occur in the mind, which is inaccessible to the exactitude of the physical sciences. Proponents of free will may be right when they claim that at least some of our volitions escape causation and are not strictly determined by the past. The trouble is that these libertarians give us no criterion for deciding which of our volitions are free in this sense and which are not. Nor do they agree as to the kinds of decisions that are likely to be exempt from causation. Thus, the French philosopher Henri Bergson, who believed that strict causation would severely limit the accomplishments of evolution, supposed that the occasions of free will are not our trivial daily choices but the decisions we make at the great crises in our lives, when our true inmost self is revealed. He thought that petty minds might pass through life without ever exercising free will.[1]

A more recent philosopher, Maurice Cranston, who cautiously conceded the possibility of free will, confessed that, when he reflected on the decisions which he made yesterday, he believed that he could have made different ones, but when he looked back ten or twenty years, he was inclined to think that he could not then have decided otherwise than he did. The decisions about which he remembered the circumstances after the lapse of a decade or two were doubtless momentous ones, affecting the subsequent course of his life, whereas yesterday's choices were probably relatively trivial, like most of those we make from day to day and soon forget. Accordingly, Cranston's view is just the opposite of Bergson's: one believed that great decisions are likely to be expressions of free will; the other, less important choices.[2]

My own experience supports Cranston's view. I would admit the possibility that when I choose between a blue shirt and a gray shirt as I dress in the morning, or when at a forking of a woodland path I turn to the right instead of the left, my choice escapes strict causation. But when I recall the crises in my life, when with much soul-searching I reached decisions that have affected its whole course, I can hardly doubt that these decisions were necessary results of my whole nature, acting in the circumstances in which I then found myself, and could hardly have been otherwise.

I do not deny that our volitions may occasionally or even frequently be uncaused, as the libertarians contend, although this cannot be proved. What worries me are the

[1] Henri Bergson, *Time and Free Will: An Essay on the Immediate Data of Consciousness*, trans. F. L. Pogson (London: George Allen and Unwin, 1910).

[2] Maurice Cranston, *Freedom: A New Analysis* (London: Longmans, Green and Co., 1953).

widespread misconceptions, by philosophers and laymen alike, of the effects that free will might have upon our conduct, and likewise of its importance for ethics and jurisprudence. Why should people flare up at the suggestion that free will is an illusion, as though one threatened to deprive them of their liberty? Probably it is the strong emotional valence of the word "free," so potent in politics, that distracts them from its special meaning in this context. Just as an appreciable amount of indeterminacy would be a source of weakness in a machine or an industrial process, so am I convinced that free will—indeterminacy in the act of willing—would be a source of weakness in human character and conduct. Accordingly, far from agreeing with those who uphold free will as essential to human dignity, I regard it as undesirable, and I would not accept it if offered to me by a god. Moreover, I am concerned that insistence upon free will prevents our recognizing, being grateful for, and using wisely the true sources of personal (as opposed to political) freedom. Before proceeding to consider these sources, let us examine the probable consequences of having free will.

As conceived by its most ardent advocates, free will is the failure of causation, the severance of the present from the past, the future from the present. Anything might follow from anything else. Could I claim as my own a volition that escapes my control and is not the necessary result of my character, acting in the prevailing circumstances? Could I have confidence in anybody whose decisions, and consequent acts, bore no relation to the constants of his nature? Doubtless nobody is wholly consistent, a failure readily explicable by the baffling predicaments in which we are often involved; but in the absence of causation our behavior might become wildly inconsistent. We could no longer trust ourselves to avoid actions that would bring us shame, disgrace, or imprisonment. One who behaved like a saint today might act like a fiend tomorrow. We would be in much the same situation as a motorist whose speeding car refused to respond to the steering wheel. If he killed a pedestrian, would the fault be his or that of the manufacturer who sold him an unreliable car?

Far from being the condition of responsibility, as moralists have held, free will would utterly destroy responsibility, and no one could be held morally accountable for what he did. In the physical world, causation, our confidence that certain results will invariably follow if all pertinent conditions are realized, safeguards our lives from day to day. It is hardly different in the moral world; without it we would be lost. We can escape from the reign of necessity only by falling into the lap of chance. Even philosophers who support the dogma of free will commonly allow us little of it, which is fortunate. To have more would be disastrous.

It is amazing that so many brilliant minds have failed to recognize that free will, the failure of causation in the act of willing, would not give us freedom to control our destiny but would leave us victims of chance. Or they have tried to make the dogma acceptable by tenuous metaphysical arguments that fail to convince. Nevertheless, the true foundation of our freedom is so obvious that one wonders how it could be overlooked. We have only to reflect upon how we reach important decisions to be convinced that, although not released from causation, our volitions are free in an important sense.

Much of the time we act from habit, which is the mechanical repetition of past actions. But occasionally, when we must make an important decision or are confronted by a dilemma, we carefully deliberate. We try to cast our minds into the future, to imagine the outcome of every alternative that occurs to us as well as how we shall be affected by it. We weigh foreseen advantages against foreseen disadvantages, incentives against dissuasives. We make an ideal excursion into the future and seem to give it a voice in its own becoming. We may do this repeatedly, varying the circumstances of our proposed action, before we reach a decision. To be sure, external restraints, imposed by nature or by society, may prevent the execution of our volition; but this also frequently occurs, however our decisions are reached.

Although all thoughtful adults must have deliberated in this manner dozens if not hundreds of times, I wonder how many have reflected upon the uniqueness of what they did. Mechanical systems, inanimate objects of all kinds, move in obedience to forces acting upon them, with no thought of where they are going or what they are becoming. They are pushed onward by the past rather than lured forward by the future. Their movements are highly predictable, if one can analyze the forces acting upon them. Even the more advanced animals never deliberate, as far as we can tell. Probably they have ideas, more or less clear, of where they are going or what they are trying to accomplish; but it is doubtful whether they try to imagine how the situation they want to realize will affect them; and it is even more doubtful whether they weigh alternative courses before they act. To us alone is given, as far as we know, the capacity to gaze with the mind's eye into the future, to compare alternative possibilities, and to choose the most attractive course. This is our freedom from blind determination by the past, as all things lifeless and probably most other animals are determined. It is not indeterminacy but a unique mode of determination which permits us to shape our future as nothing else under the Sun can.

Far from depending upon the failure of causation, or indeterminacy, in the manner of classical free will, this precious freedom that we enjoy would be weakened by any relaxation of causation. If our thoughts did not follow in a necessary sequence, each mental state determined by the preceding state, mental coherence would be lost and reasoning would become ineffective or impossible. If causation failed in the external world, permitting events to occur at random, to try to foresee and plan ahead would be a futile endeavor. Even if, in such a world, pleasant surprises occurred more frequently, unforeseen disasters would probably be much more common. Freedom, like everything else, is more firmly established in a world where causation reigns supreme than it could be if causation were weakened. We would be utterly bewildered in a world where causation frequently failed.

Although our unique freedom is an undeniable fact, it is readily lost by those who do not appreciate and cherish it. Deliberation requires calmness and patience; its great enemy is passion. When rage, hatred, lust, jealousy, fear, or any other violent emotion impels us to act without forethought, freedom is lost. These passions, often so difficult to subjugate, were imposed upon the human psyche in the long, hazardous course of man's evolution, because they increased the intensity of his efforts to overcome enemies or to flee from them in situations, such as occur among all wild creatures, when to act promptly is more urgent than to act thoughtfully. When violent passion plunges us into thoughtless action, we are driven by the past like some mindless thing, and we are no more free than a machine or a thrown stone. Hence the supreme importance, as from ancient times the wise have recognized, of keeping the passions under firm control.

Probably some readers will not be satisfied with freedom that is rooted in a unique, readily available mode of causation rather than in its occasional, unpredictable abeyance. They may fear that their efforts, toils, and pains count for naught in a strictly determinate world, where everything will be what it must be. But consider! A problem in arithmetic has only one correct answer, strictly determined by the numbers involved. Perhaps a supermathematician could predict the answer by simple inspection of the figures, but (in the absence of a calculating machine) we can learn it only by laborious addition, subtraction, multiplication, or division, as the case may be. If to know the answer is important to us, who would say that our effort was wasted because, if we made no mistake, it could not have been different?

Similarly, when we face a momentous decision, one who knows us well can often predict, with fair accuracy, what the outcome will be. But we, who imagine ourselves able to take some different course, and who perhaps feel that it is our duty to do so, must ponder and worry, make plans that we find ourselves unable to follow, go through agonies of indecision, before finally taking the course that our friend knew in advance that we would take. Would anyone assert that our painful deliberation was misspent because the result was predetermined by all the pertinent circumstances and could not have been different? Such laboriously reached decisions appear to play an important role in the formation of our characters and our attainment of happiness or success in our greatest undertakings. Without our unique method of determination, creation could never reach such high levels as with us it can. Our devoted efforts, our hopes and fears, our toils and sufferings, are not wasted but indispensable to creative advance.

Some thinkers, like Bergson, believe that in a world governed by strict causation evolution would be hampered, that it could never progress as far as it could in a world where the abeyance of causation permitted the emergence of radical novelties. This is like supposing that by adding a few letters to the alphabet we could increase the quality of literature. Creative advance depends primarily on the arrangement of the ninety-two naturally occurring elements in ever more ample, harmoniously integrated patterns. With trillions of atoms awaiting suitable conditions to join in higher syntheses, our world already contains such infinite potentialities that the addition of more would pass unnoticed.

Although our freedom depends primarily upon the mind's ability to make ideal excursions into the future and to choose among alternative courses, the body's contributions are not negligible. Few other animals have such great physiological adaptability as man. Many species are strictly confined to a narrow life zone, such as a certain altitudinal belt in mountainous country, beyond which they do not thrive. Even as individuals, we can adapt to an amazing diversity of climates, to lowlands and highlands, to arid deserts and dripping rain forests, to the tropics or the Arctic; and this does not depend wholly on our ability to prepare artificial environments for ourselves or to swathe our bodies in clothing. This physiological adaptability gives us great freedom to choose the habitat most congenial to us— a freedom which, unhappily, political and economic restraints often sadly abridge. Another contribution which our bodies make to our freedom involves the versatility of our executive organs, especially our nimble hands, which, guided by a flexible mind, can perform a vast number of operations. This permits us to choose, among many possible occupations, those most agreeable to us. Were our limbs as narrowly specialized for one or a few activities as those of many animals are, this freedom would be denied us.

Still another precious freedom is given by the adaptability of our digestive apparatus. Man is potentially omnivorous; perhaps no other species that ever existed has consumed such an immense diversity of organisms. A vast choice of diets, adequate for health and strength, is available to us. This permits us to select our foods with regard to moral, economic, and environmental considerations. As we grow more thoughtful and concerned, things which taste good to the palate no longer taste good to the mind; they may become positively disgusting. Compassion makes some refrain from consuming the flesh of slaughtered animals. Concern for the environment causes others to choose foods that can be produced with the least drain upon the land. In past years, conscientious people refused to use sugar and other articles grown with slave labor. Others avoided unnecessary beverages grown upon land that might yield food for hungry people. For all but the very wealthy, the prices of foodstuffs influence choice. Even with some of these limitations or all together, so many wholesome foods are now available that it is seldom difficult to plan an adequate diet.

It is evident that our freedom has varied sources. A unique mode of causation liberates us from blind determination by the past and permits us to set our course toward an ideal future. A flexible physiology, versatile limbs, and an accommodating digestive system enable our bodies, within wide limits, to support our choices. But how many appreciate these rare advantages as they deserve to be; how many are grateful for them? How many use them fully and wisely to lead healthier and happier lives, in greater harmony with the life around them and with the planet that supports them? All too often, men still in bondage to appetites and passions that prevent the enjoyment of their natural freedom shout and fight for greater civil liberty, difficult to win and hard to preserve. Even under the rule of an outrageous tyrant, the wise, self-controlled person can be freer than the citizen of a liberal democracy who lives in slavery to his own strong passions.

If we can dominate these passions, we become free in a more profound sense. Although our ability to make an ideal excursion into the future liberates us from the kind of causation that operates in mechanical systems and probably in the vast majority of animals, it does not insure that our volitions spring from our true and inmost selves; deliberation follows much the same course whether we plan a beneficent act or a calculated atrocity. Only volitions that express our primary nature can be considered free in the deepest sense.

Our primary nature is determined by the process that shapes our bodies and lays the foundations of our minds. This process is harmonization, which creates every living thing by joining molecule to molecule and cell to cell to build its organs and which unites these organs in an integrated system that insures the harmonious functioning of a healthy organism. Likewise, it impels us to strive for that coherence among the contents of our minds which is our most reliable criterion of truth. An animal formed in body and mind by such a process should be consistently pacific and creative, seeking above all to dwell in harmony with all its neighbors, of whatever kind.

Unfortunately, this primary nature is overlaid, more or less deeply according to the species and the individual, by appetites and passions, often uncontrollably strong, that have evolved to promote survival of the individual and perpetuation of the species in a competitive, overcrowded world. These accretions often so thoroughly mask our primary nature that certain psychologists have concluded that they constitute the essential ego; but, if they had delved more deeply into the human psyche, they might have discovered something still more fundamental. If aggressiveness, hatred, lust, and similar characters were expressions of our central selves, it should follow that, if they could be extirpated, little would remain. But what do we actually find? Those most successful in disburdening their spirits of these distressing accretions most emanate love and goodwill, strive most earnestly to cultivate harmony with all around them. This, as we would expect, is the true expression of the inmost nature of a being formed by harmonization. And what could be freer than a volition determined by what we essentially are?

Let us now follow the genesis of a volition that is free in the most profound sense. Its motive springs from our inmost nature and conforms to the process of harmonization that created us. Before willing to act, we deliberate, trying to foresee the outcome of various alternative courses; and we prefer that which promises best to preserve or increase harmony, in our own person and in our relations with our surroundings, living and lifeless. Our volition is determined by our primary nature in relation to the opportunities for its full expression, or the obstacles to such expression, that we detect in the external world. The failure of causation—classical free will—at any point in the complex internal process of choosing a course of action with momentous consequences to self or others, far from increasing our freedom, might, by permitting the intrusion of chance, make our volition less than faithful to our inmost nature, whence our motive springs.

No matter how straitened our circumstances or how hostile the world around us, nothing outside ourselves can completely annul our freedom without first debilitating our minds, as by drugs or prolonged enforced sleeplessness. So long as the mind is unimpaired, two courses are always open to us: to act or to refuse to act. In the clutches of a fiendish dictator, who offers the alternatives of abetting nefarious schemes or suffering tortures, the strong-willed person can remain true to his principles while his body is horribly wracked. Although we lose our freedom whenever uncontrollable passion precipitates us into thoughtless action, nothing external to ourselves can so thoroughly deprive us of it.

12. Man's Compassion and Nature's Harshness

Every development that binds creatures closer together, increasing harmony and mitigating strife, is a momentous step in life's forward march. Among these developments are beauty, cooperation, and compassion. Cooperation may take the form of a united effort in a common endeavor, as in repelling an enemy or building a nest. Or it may be an exchange of benefits, as when plants supply nectar to the animals that pollinate them or offer fruits to those that scatter their seeds, or when cleaner fishes derive nourishment by removing parasites and dead tissues from client fishes. Although the evolution of some of these associations is complex, it can be explained by widely accepted evolutionary theory, for cooperation helps both parties survive and reproduce.

Compassion is in a different category. Far from increasing the biological fitness of the compassionate one, it often entails a sacrifice, as when he refrains from destroying an animal that might yield food, is lenient with those that menace his crops or property, or devotes time and resources to alleviate the suffering of those who make no return. To current evolutionary theory, compassion is an anomaly. It arises when an animal reaches a certain critical point in its psychic ascent. The day when any creature, human or otherwise, gave evidence that it had attained this point, as by desisting from killing some intended victim because of fellow feeling or going out of its way to feed the hungry, was certainly one of the most sublime in life's long advance.

When we realize how anomalous in the living world compassion is, we understand why those who treat nonhuman animals with mercy and forbearance, and who sometimes try gently to persuade others to do likewise, are often regarded with amused tolerance if not with scorn by the tough-minded. These critics of one of the most admirable of human attributes remind us that nature is "cruel," that merciless competition is nature's law, that everywhere the strong take what they want with no consideration for the feelings of the weak, that man is part of nature, and that we must conform to nature's way. Compassion, they may assert, has its place in the relations of man to man, but it is hardly applicable to man's dealings with the rest of the animate creation.

This alleged cruelty of nature deserves our most careful examination, for our whole attitude toward the living world will be profoundly affected by our view of it. In general, philosophers, who cast a coldly appraising eye upon nature, have taken a rather gloomy view of its harshness. One recalls Herbert Spencer's opinion that torturing parasites outnumber in their kinds all other organisms,[1] as well as Schopenhauer's vivid accounts of the sufferings of life. But naturalists, who gratefully recall many happy hours in the woods and fields, incline toward the opposite view. They tend to emphasize the joys and satisfactions of free animals, to minimize their sufferings, and to brush aside the disagreeable facts of the natural world. Probably the truth lies somewhere between these extreme attitudes.

Before we try to settle this question, we must have clearly in mind what we mean by "nature" and what by "cruelty." A characterization applicable to nature regarded as a personal being directing the affairs of Earth and its inhabitants may be inappropriate if we hold some other concept of nature. In the present context, we mean by nature the totality of the processes that have created and that maintain the physical world and all the living things it supports.

---

[1] Herbert Spencer, *The Principles of Ethics* (New York: D. Appleton and Co., 1896), vol. 2, p. 5.

Nature may be something more than this impersonal aggregate, but, for our immediate purposes, this concept will be adequate. Turning now to cruelty, it is necessary to distinguish clearly between callousness and the infliction of suffering for the pleasure or amusement it affords the spectator. The carter who mercilessly goads his overloaded oxen is probably callous rather than cruel in the narrow sense of the word. His motive is to get his cart up the hill rather than to make his animals suffer. On the other hand, the boy who tortures an animal to see it writhe or the spectators who gloat over a bloody bullfight are cruel at heart.

Of cruelty in this narrow sense, there appears to be little in the nonhuman world. The cat's play with a mouse is a favorite example of the innate cruelty of animals; but I doubt whether it finds pleasure in its captive's fear and pain or even thinks about it. The skillful execution of a stereotyped act seems to be the source of the cat's gratification; the mouse's feelings probably do not enter into its picture of the world. The majority of predatory animals appear to capture no more victims than they need to satisfy their hunger and feed their young, and they seem to kill with no deliberate elaboration of the act. The death of creatures devoured by predators, horrible though it be, is probably, in most cases, much less agonizing than that of the millions of unretrieved victims of gunshot that hunters each year leave to the misery of festering wounds and slow starvation. Nevertheless, the death of animate prey is not always as sudden and merciful as we like to imagine it. Hawks sometimes calmly proceed to pluck all the feathers from a captive bird that is far from dead; and carnivorous animals, no less than savage men, at times tear the flesh from a living victim.

If, to the countless millions of animals of land and sea that daily surrender their lives to fill the maws of other animals, we add the myriads more that die slowly and horribly from bacterial, protozoan, and viral infections, as well as all those that are gnawed, punctured, and lacerated by an innumerable horde of parasites of the most diverse sorts, I believe we must agree that the amount of carnage and mutilation in nature is incalculably vast. How much pain and suffering accompany all this disease and death is another question, one far more difficult to answer. As every amateur in skepticism knows, pain, like pleasure, is strictly demonstrable only to the person who feels it in himself. It is just here that the naturalist-writers who like to hide nature's disagreeable aspects try to profit by our ignorance. They cannot deny the very obvious fact that there is much carnage in nature, but they sometimes maintain that it is accompanied by much less suffering than we naïvely imagine, citing, perhaps, the well-known story of the African explorer David Livingstone, who felt no pain while actually in the jaws of a lion.

It is most difficult for us to assess the joys or the sufferings of beings, including people of alien cultures, whose heredity and background are different from ours. At times I have been amazed by the number of kicks, each hard enough to stun a man, that a horse will take without seeming to mind them, yet a horse with a disordered stomach gives every indication of the most acute distress. But since suffering and happiness are, at least in ourselves, correlative states of consciousness, if we diminish the capacity for experiencing one we must, in all consistency, reduce in the same measure the capacity for the other. If animals can be lacerated and killed without feeling much pain, it would seem that they must live without much pleasure or contentment (as suggested in chapter 6, this may not apply to insects). We are reduced to the Cartesian doctrine of animal automatism or something closely approaching it. Thereby we divest the pageant of animal life of most of its significance—and this is just what the naturalist-writers, on the whole, seek to avoid.

For my part, I frankly admit that the animate world, from the lowest protozoan up to the highest vertebrate, is full of strife and carnage, which are, to all appearances, accompanied by a volume of suffering inconceivably great. To me, nature is always interesting and often beautiful; but, at the same time, it has an aspect so terrible that many of those who ecstatically contemplate it through a roseate haze could hardly bear the vision if the mist were suddenly blown away, to reveal the natural world in its stark nakedness. Those who write or teach about nature should regard it as a duty to point out its terror no less than its beauty, beneficence, and interest. Only so can they guard against the future bitter disillusion of those whom they influence.

But this harshness of nature, far from invalidating human compassion, gives it all its significance and grandeur. What would be the use of compassion in a world so ordered that strife could not arise among its varied inhabitants, in a world where pain and misery could never occur? It is just the presence of suffering that imparts value to compassion; and the more strife and pain the world contains, the more precious compassion becomes. It will be recalled that Buddhism, often called the religion of pity, never denies the fact of suffering. Quite the contrary, its whole doctrine is built upon "the truth of suffering," the first of the Four Noble Truths proclaimed by its founder in his first discourse to his disciples. Just as an optimistic philosophy is but a flimsy doctrine if it has not squarely faced and assessed all the disagreeable truths that pessimism proclaims, so a compassionate attitude toward the living world is of little worth if it deludes itself as to the magnitude of the world's suffering.

What, then, is the origin of this compassion, this tender forbearance, which, as far as we can be sure, shines forth only here and there in a world of conflict and pain, like a feeble ray of light in the blackness of a vast subterranean cavern? Only two alternative explanations seem possible: either it is a development of nature itself, or it was implanted in man by some agent that stands above and beyond nature. If the latter, then it would appear to possess the highest possible authority, so that to disregard or disparage

175

this sentiment within us would be sinful and irreligious. But to take the contrary view, and regard compassion as a natural development, by no means divests it of sanctity and a claim to our respect.

Everywhere in the natural world we detect compensatory trends, which prevent the upsetting of the natural order by an unopposed process and preserve the balance of the whole. Thus, the more we compress a fixed amount of gas, the more strongly it resists the force applied to it. By becoming denser, the air in which we live at the surface is enabled to resist and support the very considerable weight of the atmosphere above it. As a wire is heated by an electric current, it offers greater resistance to the current's flow, thereby reducing the probability that it will become so hot that it melts. The faster a body moves, the more it resists further acceleration; to double the velocity of a moving mass requires four times the energy that was needed to give it the initial velocity from a state of rest. A compensatory trend with momentous consequences for our planet's climate and its life involves the expansion of water as it approaches the freezing point, which reduces the probability that lakes and seas will freeze solidly, as explained in chapter 3. The widespread occurrence of compensatory trends in physicochemical systems is recognized by the theorem of Le Châtelier, which states that, if we bring an additional force to bear upon a system in equilibrium, the point of equilibrium will shift in such a direction as to diminish the effect of this force.

In the growth of compassion, we have an example of an analogous process in the living world, which is essentially a system in dynamic equilibrium. On every side, we behold the uninhibited exploitation by living things of other living things, with no consideration for the feelings or purposes of these exploited creatures. But the exploiters have, at best, a low degree of intelligence, so that they can use these other creatures only in certain ways determined by their own hereditary organization; and, in many instances, the relations between exploiter and exploited, between predator and prey, have, through countless generations of mutual interaction, reached an equilibrium that permits the continued prosperity of both, as species if not as individuals. But, after long ages, a new force springs up in the world, in the form of an animal far more intelligent than any that preceded it—an animal able to devise countless novel, ingenious, and often diabolic ways of exploiting its fellow animals. The equilibrium between the forms of life, the whole system of nature, might be overturned or utterly destroyed by this more cunning animal, if a new factor did not come into play as a principle of limitation.

The very intelligence that makes this animal so much more efficient as an exploiter admonishes it that it is wrong to press its advantage over other creatures to the utmost limit. Its spontaneous feelings rise in revolt against the merciless exploitation of other animals. An inhibition springs up from the inmost depths of this intelligent animal and tempers its cunning with mercy. Compassion is born. This newly springing sentiment seems to be nature's effort to mitigate the strife that results from its own teeming fecundity, to outgrow the crude methods of the primal ages and bring a milder dispensation upon Earth. And if, in the manner that I have perhaps too hastily sketched, we view compassion as a higher development of a natural process, to which all the preceding stages of life are foundational and preliminary, it seems just as sinful and recalcitrant to scorn or smother the first faint glimmer of this new light in our breast, as when we believe that it was implanted there by some higher power.

For a more detailed account of the natural origin of compassion and allied sentiments, I recommend for the interested reader a voluminous work by the Australian naturalist-philosopher, Alexander Sutherland. In *The Origin and Growth of the Moral Instinct*, Sutherland traced in great detail the development of moral attitudes from the relation of the parent animal to its young. Parental sympathy expands until, in modified form, it embraces not only unrelated individuals of the same species but, at last, those of different species; and compassion is a product of this sympathy. Thus, we seem to detect a sort of rudimentary compassion at work in the behavior of birds that feed, and sometimes even brood, the nestlings of other parents, perhaps of distinct species, and in that of mammals who suckle the young of some other animal. But a developed compassion, conscious of its purposes, seems possible only where imagination is more active than it appears to be in non-human animals. Compassion is the flowering of a highly endowed mind, as cruelty is its hideous perversion. As only a rational animal can talk nonsense, so only an animal capable of compassion can be deliberately cruel.

One further question sometimes troubles those who, viewing with compassion the multitudinous creatures around them, are eager to do what little they can to ease the pains and increase the joys of all things that share with them the boon and the burden of life. When we contemplate the exceedingly complex interrelations among living things, we often wonder whether some intended act of kindness will not, on the whole, increase rather than diminish their pangs. We see an ant drowning in a puddle and are tempted, like the dove in the fable, to offer it a straw to which it can cling and be rescued. But we reflect that many ants are predatory and, if this one continues to live, it may kill other tiny creatures, so that in saving its life we may doom other beings to death. Or we hesitate to brush aside the web that a spider has spun across our path, yet we recall that, if we walk around instead of pushing through it, many a hapless insect will fall into its snare. Or we are moved to return a fallen birdling to its nest but remember that it will devour many an insect and many a worm. On the other hand, recalling a speculation of Erasmus Darwin, we might argue that by killing some large animal, even a man, we actually increase the sum of pleasure in the world, by providing subsistence for the innumerable maggots and other organisms that batten in the carcass.

To attempt to apply such a Benthamitic calculus to the pleasures and pains of the whole creation can lead us only into a hopeless muddle. It is most doubtful whether such a summation of pleasures is possible even within the far narrower bounds of a human society. By what scale shall we measure the delights of a dining spider or the agonies of an ensnared moth? All that the compassionate person can hope to do is to examine his acts in their more immediate consequences, without groping to follow their repercussions to the remotest shores of the living world. The question he must ask is not "What can I do to diminish the sum of pain among living things?" but "What can I do to reduce the suffering for which I am directly responsible or which comes immediately to my attention?" When we help some animal in trouble or refrain from some course that injures living things, we cultivate and satisfy a sacred impulse within us; when we take the contrary course, we thwart and violate this impulse. It is not our fault if the world is so constituted in its multitudinous interactions that nothing we can do will lessen by a single twinge the sum of its pains or add a gleam to the total of its happiness. Perhaps fortunately for our finest sentiments, we can never know this of a certainty. We can only hope that, by lightening a burden here and providing a small joy there, the balance will be favorably altered, as in all probability it is.

But it is not so much by our positive acts as by our restraint that we humans most benefit other creatures. It does not follow from this that the person who does most for the living world as a whole is the most passive and indolent—to advocate such a doctrine of inactivity would be perverse. As Gandhi taught, only the truly courageous man can succeed in the practice of nonviolence. To desist from some cruel amusement in which all one's acquaintances indulge, to stop using some common article because its production entails suffering by men or animals—this requires a measure of fortitude far from commonplace. To simplify our lives, to find methods of satisfying our basic needs with less destruction of others' lives, demands careful thought and often, too, an active inventiveness.

If I have touched in this chapter on the sufferings of the living world, it is not because I believe it wise or wholesome to spend much time brooding over them. The great systems of spiritual culture have generally had scant use for pity. They have exhorted men to bear with equanimity their own disappointments and inevitable pains; and what would we gain by this strenuous self-discipline if we were to be upset by all the misery that we can hardly avoid seeing when we look around us? Such sights would agitate that calm, serene mind that the wise have always striven to cultivate. Compassion is still at root a passion, a passive affection of the mind; and, from Buddha to Spinoza, the great doctors of the spirit have recommended the subjugation of the passions.

But when, instead of brooding over suffering, we do something to alleviate or remove it, we are no longer passive but active; and such activity befits a mind that is both noble and calm. This was the course of the Stoics, who deprecated pity yet cultivated the most effective philanthropy of the Classical world and were largely responsible for the "golden age" of the Antonines. It is far better and more worthy of us to perform one smallest act of kindness than to spend a lifetime brooding inertly over the woes of all the world. And by such activity we diminish the distress that the sight of suffering causes us, no less than the suffering itself. The wisest course is to pay attention to just so much of the world's distress as we can hope to alleviate. To distract our spirits with the remainder is futile.

Thus, when those of us who feel compassion for nonhuman creatures are reminded, sadly or mockingly, that nature is "cruel," let us freely grant the contention. We may even permit the hard-bitten "realists" to paint the harshness of nature in the blackest colors that they can find;

their portrait cannot alter the quality of the original that it purports to represent. But it will be useless for them to deny the reality of such compassion; we have felt it in our own spirits, and we have excellent evidence that other men have felt it for at least three thousand years and probably much longer.

Then let us ask our critics whether they regard this compassion as a natural evolutionary development or as a sentiment implanted in the human mind by a power that stands above nature and directs its course. If the latter, then they have invested this sentiment with the very highest sanctity, and even to speak disparagingly of it is impiety. If the former, then we must see in compassion an effort by nature to transcend the crudities inseparable from the earlier stages of life's advance; and as such, too, it would appear to be authoritative as one of the latest and highest products of the aeonian movement that made us what we are. If this sentiment is still weak and sporadic, we recall that most great things had small and unpromising beginnings; and, against the background of possibly 3,000 million years of life on this planet, compassion is still in its infancy, and it should grow stronger as life ascends to higher levels.

13. Conservation, Man's Debt to His Planet

 For thousands of millions of years before man appeared on Earth, our planet was preparing for him. Not only was it becoming able to support a large animal with varied physical needs, it was also becoming a place to stimulate his unfolding spirit. By the beauty and majesty that surrounded him, it fostered his aesthetic appreciation, aroused his sense of wonder, and made him grateful for the boon of life in so fair a scene. A more austerely plain environment might have supplied man's physical needs; but in it would he have become capable of producing poetry and art, of imagining all the charming unseen beings that adorn mythology, of developing the philosophies and sciences that are born of wonder—would he, in short, have become fully human, in the best sense of the word? We are what we are because we had the good fortune to evolve on a planet well fitted to nurture the spirit as well as the body.

Earth's Twofold Fitness for Man

This twofold fitness of Earth for human life must be kept in mind in all our efforts to conserve its resources. If we neglect one aspect of the problem, we may starve physically; if we neglect the other, we may starve spiritually; and total neglect could doom mankind to waste away in both body and mind.

Up to a certain point, man's alteration of his environment increases both aspects of its fitness to support him. Evolution brings diversification, enriching the Earth with variety; human activities have accelerated this process. A landscape pleasantly varied, with attractive homes standing in lovely gardens amid fertile cultivated fields, interspersed with tracts of unspoiled woodland, can be as aesthetically satisfying as a vast tract of unbroken forest; and in such a setting one need not go far to enjoy wild nature. But the farms should be restricted to the more fertile valleys and gentle slopes; the rugged hills should retain their protecting mantle of trees; the wetlands should remain the inviolate home of waterfowl; the streams should flow pure through woodland and tilth.

Unfortunately, this ideal balance of man and nature is difficult to preserve. As the human population increases, forest on steep slopes is destroyed to make farms; wetlands are drained; rivers lose their clarity. Already two thousand years ago, the Roman poet Lucretius described the very process that I have painfully watched in the tropical country where I live:

> And day by day they'd force the woods to move
> Still higher up the mountains, and to yield
> The place below for tilth.

This invasion of inferior lands by agriculture leads to a most unwholesome situation. Unless whole mountainsides are terraced with retaining walls, as in Peru and the Philippine Islands—an undertaking that seems too laborious for modern hill farmers—the steeply inclined fields erode, leach, and lose their fertility, especially under heavy rainfall. More and more land is needed to keep undernourished multitudes alive. The combination of deforestation and hunting destroys the woodland fauna; the pleasantly diversified landscape becomes a monotonous expanse of impoverished farms dotted with dilapidated dwellings. Meanwhile, those who can no longer make a living on the land drift to the cities, where ever swelling multitudes make huge demands upon all Earth's resources. As the mass of human flesh swells, the world has ever less to offer to the human spirit. Those who seek the beauty, tranquility, and intellectual stimulation of unspoiled nature must travel ever farther to find it. Finally, the world is faced with shortages of such basic necessities as food, water, fuel, and materials for construction.

The Origins of Conservation

Man is not unique in his tendency to ruin his environment as he becomes too numerous; without checks upon their increase, other animals do the same. Grazers and browsers destroy the vegetation that supports them wherever they consume it more rapidly than it can grow. Defoliating insects become so numerous that they kill the trees which nourish them. Carnivores can so reduce the numbers of their prey that they starve. Among the animals least likely, or able, to harm their environment are the frugivores and nectarivores, which of all creatures have achieved the best relationship with the plants that feed them. When ripe fruits are scarce, frugivorous birds may turn to devouring them green, thereby reducing the number of viable seeds that they disseminate; but long-lived trees and shrubs could survive this for years, while starvation brings the avian population into balance with their productivity. Similarly, it is difficult to imagine how hummingbirds, bees, and other nectar drinkers that are not, like butterflies, devourers of foliage at one stage of their lives could permanently harm flowering plants.

Although man is not the only animal able to wreck his environment—with his huge populations spread over all the planet and with the extremely varied demands he makes upon its bounty—he can and does devastate the environment on a scale, and with a thoroughness, equaled by no other animals except, perhaps, the vanished dinosaurs. Frequently, when some process or evolutionary trend threatens to become too violent or destructive, compensatory forces arise to restore equilibrium. The theorem of Le Châtelier recognizes this fact in the physical world. When the Mesozoic saurians became too huge and destructive, nature eliminated them. When man's capacity to inflict suffering and death upon his fellow creatures became excessive, compassion arose to mitigate his cruelty. Similarly, when we view man's concern for conservation in the broad perspective of evolution, we might regard it as nature's attempt to compensate for the excessive powers of reproduction and the acquisitiveness that it gave him. Animals less able to destroy their environment are held within safe limits by other means; conscious concern for the environment and the varied life that it supports appears to be peculiar to man. A recent development at the higher levels of evolution, it is still in its infancy; whether it will grow strong enough to save our planet before teeming billions of humans overwhelm it remains to be seen.

Like other evolutionary developments, concern for the environment first arises in individuals who appear as unpredictably as a biological mutation. Whether actual genic mutations are involved is doubtful; more probably, the concerned individual has received from his parents a unique combination of their genes, which, together with cultural influences, has given him a broader outlook and a heightened sense of responsibility. Here and there, some thoughtful person awakes with alarm to what is happening to the natural world and does what he can to save it. Usually he is concerned with the particular aspect of the deteriorating situation that is closest to his interests. Agriculturists are disturbed by the erosion and decreasing fertility of the soil; wilderness lovers are distressed by the shrinking of the forests; hunters complain of declining numbers of shootable animals; bird lovers lament the destruction of the feathered creatures that delight them; fishermen bewail the pollution of streams and the diminishing rewards of patient angling; biologists deplore the disappearance of species that they wish to study; the growing scarcity of wildflowers perturbs others. Those of similar interests band together to save what most concerns them. Of such dissimilar, and often conflicting, interests the modern conservation movement was born.

Finally, those of widest vision recognize that the natural

world of soil and water, vegetation and fauna, is one vast network of interacting parts, which must be preserved as a whole, lest it collapse as a whole. To succeed, conservation must become a unitary endeavor to save the environment, enlisting the cooperation of all the special interests with which it began. It must give equal attention to both aspects of Earth's twofold fitness to nurture man. If it makes no effort to conserve Earth's beauty and the living creatures that share it with us, it will lose the allegiance of all those whose spirits revolt at the prospect of being imprisoned in a narrowly human world. If it neglects the economic aspect, not only will it fail to preserve the natural foundations of civilization, but it will soon find itself unable to protect from the hungry multitudes any scraps of wilderness that might yield a meager return to cultivation or any animals or plants that might be eaten or otherwise used. A sound policy of conservation must be equally sensitive to economics and aesthetics.

The earliest European colonists in North America found themselves on the eastern edge of a vast continent whose immense resources had hardly been touched by the sparse aboriginal population. Pushing ever westward with admirable energy but deplorable lack of vision, they proceeded to exploit the continent as though its bounty were inexhaustible. Only after the world's most extensive broad-leaved, temperate-zone forest had been reduced to a few small stands of mature trees, after some of the most abundant mammals and birds had been brought, in cases irrevocably, to the verge of extinction, after farms had been exhausted by careless husbandry and great rivers foully polluted, did a few alert individuals become sufficiently concerned about all this heedless destruction to arouse their more intelligent neighbors and finally persuade their governments to take protective measures. In the white man's North America, concern for the environment, now happily so widespread and growing, is hardly a century old.

Actually, on the wider view, such concern proves to be

very old. Primitive man, unable to call for succor from some neighboring nation in time of famine or other disaster, long ago recognized the need to husband the resources of the territory controlled by his tribe or band. Animistic beliefs supported a policy that was doubtless dictated by sound insights and, perhaps, memories of hunger following careless exploitation. In widely scattered parts of the world, trees could not be felled without propitiating their indwelling spirits or providing other abodes for them. Among South American Indians, each species of hunted animal was believed to be protected by a supernatural guardian, often an outstanding individual of its kind, who would punish the hunter that killed in excess of his needs, either by frustrating his attempts to find more or in other ways. For many primitive tribes, killing animals was too serious a matter to be done for diversion, in the manner of millions of well-fed modern hunters who have no need of their victims' flesh. Moreover, these tribes commonly limited their population to the productive capacity of their territory, by continence, abortion, or infanticide. Without such limitation, all conservation measures will soon prove nugatory.[1]

The association between religious doctrines and practices that favored the conservation of nature, already evident in the animistic beliefs of primitive peoples, was carried over into certain more advanced religions. No genial dreamer has had a vision of Earth's future more alluring than that of the prophet Isaiah, who foresaw a time when absolute peace would reign throughout nature, when the wolf, no longer fiercely carnivorous, would lie amicably with the defenseless lamb, and the child could without peril play upon the hole where the asp lurked. Although, unlike the Greek philosophers, the Hebrew prophets seem rarely to have developed their views in elaborate detail, the implications of the

[1] Early man's attitudes toward plants and animals are discussed more fully in the present writer's *The Golden Core of Religion* (London: George Allen and Unwin, New York: Holt, Rinehart and Winston, 1970), chapter 6.

messianic vision are clear. Destruction of every kind would cease throughout the natural world, thereby realizing the conservationist's fondest dreams with a thoroughness that he cannot rationally hope to achieve. Yet messianic peace seems to be the logical end of harmonization. The movement that binds many organs, many millions of cells, and countless billions of atoms into a harmoniously functioning unity in the body of a larger plant or animal would, we might suppose, eventually unite all living things in a single society of harmoniously interacting individuals, cooperating together for the good of the whole as closely as do the organs of a healthy body.

Although the words of the Hebrew prophets are treasured to this day, they appear to have had, at best, only a minor influence on the practices of their contemporaries, including the priests who controlled the national religion. It is not evident that any measures were taken toward even a partial realization of a generous vision that is not likely to be fulfilled within the foreseeable future. Despite the denunciations of the prophets, who declared over and over that Yahweh demanded righteousness, justice, and mercy rather than burnt offerings, the bloody sacrifices continued until the Romans, in A.D. 70, destroyed the temple in Jerusalem, where alone they could be performed.

Meanwhile, India was the scene of an ethical and religious movement that yielded the first comprehensive conservation legislation of which we have knowledge, among the most inclusive that has ever been enacted by any government. After the Emperor Asoka Maurya embraced Buddhism, about the year 261 B.C., he stopped his wars of conquest, renounced the chase, to which he had been addicted, and administered his extensive empire in accordance with the teachings of the "religion of pity," without, however, interfering with other faiths—which this very liberal monarch actually supported. On elaborate columns, boulders, and outcropping rocks throughout the empire were carved the royal edicts, which tell a remote posterity that Asoka

prohibited the killing of animals so diverse as parrots, star-
lings, geese, doves, and other birds; of bats, tortoises, river
skates, boneless fish, and queen ants; of porcupines, tree
squirrels, barasingha deer, monkeys, and rhinoceroses, as
well as all other quadrupeds that were neither utilized nor
eaten. For the kinds of fish whose capture and eating were
permitted, closed seasons were established; on certain spe-
cified days, fish could be neither sold nor eaten. On these
same days, the destruction of animals of any kind in fish
ponds and elephant forests was forbidden. Setting fire to
forests, either wantonly or to drive out hunted animals, was
prohibited. Chaff from threshing floors could not be burned,
because of the small living creatures that lurked in it. The
castration and branding of domestic animals were controlled. Female goats, sheep, and pigs were exempted from
slaughter while they were with young or in milk, as were
their offspring up to the age of six months.[2]

In ancient China, Taoist piety protected even insects and
herbs from wanton destruction. The only firm foundation
for conservation is an attitude toward Earth and its productions that is essentially religious, whether or not it regards
these things as directly created by God. Under the influence of their religions, men have commonly cultivated
longer thoughts than when motivated primarily by economics, in which concern for the future is always precariously pitted against the desire for large immediate profits.
To be sure, conservation in the modern world must lean
heavily upon science and technology. Agriculturists and
ecologists must indicate which lands are most suitable for
agriculture; foresters must manage woodlands for sustained
yields; ornithologists are best able to point out the conditions in which birds flourish; and hydrologists should guard
our water resources. But what scientists and technologists
accomplish depends largely upon their motivation and the

[2] Vincent A. Smith, C.I.E., *Asoka: The Buddhist Emperor of India*,
3d ed. (Oxford: Clarendon Press, 1920).

attitudes of the wider public that supports them. What attitudes or mental attributes are needed to insure the continuing protection of a natural world that is under constant pressure from the many who would ruthlessly pillage it?

Attributes of the Conservationist

Above all, grateful appreciation of Earth's beauty and fruitfulness prompts us to preserve the environment from destructive exploitation. True gratitude is one of the rarest and most gracious of virtues, confined to highly organized minds whose thoughts pass swiftly from benefit to benefactor, and who would sooner forgo the gift than harm the giver. Only the thoughtless person can fail to recognize that the ultimate foundation of everything he cherishes is the natural world, and, if capable of gratitude, he will stubbornly defend it.

Out of gratitude grows love. The grateful man loves the natural world for its endless beneficence to himself and those dear to him. Mere utility is not an adequate foundation for love. A tool or machine may be useful to us, yet we do not truly love it, for if it wears out while serving us we discard it like a tattered garment. An object of enduring love must be more than useful and contain more than it gives us; it must have a vast reserve of lovable qualities which cannot be exhausted by the benefits it lavishes upon us; it must transcend our grasp, like a loved person. This is preeminently true of nature, which, unless relentlessly plundered by hordes of hungry or greedy men, will remain an inexhaustible store of beauty, interest, and material benefits, nourishing and sustaining present and future generations in body and spirit, yet ever remaining too great and mysterious to be seized entire by man. Not only the whole of nature but its separate components, vegetable and animal, evoke the love of those who zealously protect them.

The incredible cruelty with which many workers in the life sciences treat sensitive animals is one of the most shocking manifestations of our time. Without compassion, the conservationist can become hardly less callous than the wanton exploiter. If he directs conservation programs, he perforce deals with populations of animals rather than individuals. Occupied with statistics and administrative duties instead of intimate contacts with living animals, he too easily forgets that each of the latter is a sensitive creature clinging to its life, able to enjoy and to suffer. All his measures should be directed to preserving conditions favorable to them; even when, to prevent a disastrous disequilibrium in a natural community, he must consent to the destruction of certain of its members, he should do so with pitying reluctance.

It is a curious fact that prominent among the early conservationists in the Western world were hunters disturbed by the threatened disappearance of the animals they pursued; not compassion but the continuance of their pastime was the motive that prompted their efforts to enact protective legislation and establish refuges for wildfowl and other animals. Because they were many, well organized, and often wealthy, such foresighted "sportsmen" were, and are, powerful allies of the conservation movement. Yet their activities always distress true friends of the living world, who can never forget that a large proportion of the gunners' victims, denied the mercy of sudden death, die lingeringly of wounds or prolong a crippled existence. Conservationists motivated by gratitude, love, and compassion hope that eventually they will become strong enough to accomplish their objectives without such unfeeling and essentially selfish allies.

Among conservationists two attitudes prevail, which, by analogy with governments, we may distinguish as totalitarian and liberal. The totalitarian conservationist is concerned with keeping the whole natural community running smoothly or in ecological balance, as he would say; as long as this can be done, he is not greatly troubled by how harshly the system falls upon individual animals. He is not, in principle, opposed to hunting and trapping, so long as these cruel practices do not threaten to exterminate a species or upset nature's balance. The liberal conservationist desires each living thing to enjoy all the happiness or freedom from stress of which it is capable. Aware that an animal does not cling to its life the less because its kind is in no danger of extinction, that an individual of an abundant species is as anxious to live as any of the few dozen survivors of an endangered species, he is compassionate and deplores killing, even when it does not jeopardize a species or impair the ecological balance. Although he knows that unless the whole natural community remains in a flourishing state individual organisms will suffer, for him the community exists for the welfare of the individuals, rather than the reverse, just as, for a liberal statesman, the country must be kept prosperous for the welfare of its people. The totalitarian conservationist is like the totalitarian ruler who is not averse to oppressing the people so long as the country remains rich and powerful.

Not least among the motives of the zealous protector of the environment is generosity. He wishes others to share the benefits that he receives, to enjoy what he enjoys, not only the material means of subsistence but all the treasures of beauty and knowledge that nature offers to those capable of appreciating them. His generosity extends not only to his contemporaries but to all future generations, not only to other men but to the living world as a whole.

"Be thrifty so that you may be generous" was an old Taoist maxim. By no possible economy can nature provide an abundance for all the living things that it engenders. Everything that we eat, every natural product that we use, is in effect something taken from some other creature that might profit by it, now or in the future. One contribution that everyone can make to the cause of conservation, no matter where he dwells, is to waste nothing, neither food, nor clothing, nor paper, nor fuel, nor any other of Earth's products. The inordinate waste of prosperous modern societies immeasurably increases the drain upon natural resources and needlessly accelerates the spoliation of the environment. The old-fashioned virtue of frugality must become the central preoccupation of the environmentalist whose dedication to the cause is more than nominal. Too many people, delighting in the display of their opulence, deny themselves the satisfaction that comes from making things yield their last ounce of usefulness, without depriving themselves of food or other necessities.

The frugality that supports our generosity need not apply to the life of the spirit, which is commonly exalted rather than depressed by a mild asceticism. The food that I eat cannot nourish another, and the garment that I wear does not keep another warm; but I can share my knowledge and insights with countless others without diminishing my store of them, and I can enjoy beauty yet leave it undiminished for the delectation of others. The economy of the spiritual world is exempt from the harsh limitations that cramp the material world. In this realm, we need never stint ourselves lest we deprive others; on the contrary, by sharing the treasures of our minds we often take firmer hold of them.

One of the most necessary attributes of the conservationist or environmentalist is foresight, the habit of looking to the remote as well as to the more immediate consequences of all his activities that in any way affect the natural world. Concerned for the welfare of generations yet unborn, he will not readily accept present advantages to the detriment of Earth's future productivity. Moreover, he will need courage to resist the threats and ridicule of the ignorant, the selfish, and the greedy, who cannot understand or appreciate his willingness to make sacrifices for that which yields him no direct personal advantage.

The conservationist must approach his decisions and tasks with humility. Humbly recognizing that nature is too vast and complex to be fully understood, he will hesitate to interfere with natural processes or upset natural balances, lest the outcome of shortsighted meddling be far other than he anticipates. He must avoid the hollow pride or smug satisfaction that asserts that man is the end and purpose of nature or the single branch of the animal kingdom that can continue indefinitely to evolve to higher levels. His humility will make him suspect that the living world has potentialities that are hidden from him, that it realizes values of which he is unaware. He will try not to subject the whole planet to man but to fit man harmoniously into the living world.

Finally, we need reverence, not only the "reverence for life" that Albert Schweitzer taught us but also reverence for the creative power hidden at the heart of the Universe, of which life is an expression. In the natural world we are in the presence of forces immeasurably greater than ourselves, which at best we only imperfectly understand. Not power but goodness deserves our reverence. Since these forces have brought forth beauty, joy, and man's conscience, they must be regarded as on the whole beneficent, even if, in the immense complications of the living world, they have secondarily caused much conflict and suffering. In our efforts to protect nature, we are dealing with these forces and their creations. The attributes or attitudes that we need to pursue our endeavor with some prospect of success differ little from the virtues taught by advanced religions and religious philosophies. Unless a substantial fraction of mankind comes to regard nature with an attitude essentially religious, including gratitude, love, compassion, generosity, humility, and reverence, and becomes frugal in the use of its bounty, all the devoted efforts of a handful of earnest conservationists and environmentalists are doomed to fail.

Conservation and the Unity of Mankind

Reverent regard for the planet that bears us and steady determination to keep it fruitful and beautiful are our best hope for the unity of mankind. Although for millennia prophets and philosophers have taught that all men, of whatever race or color, are brothers, the concept of brotherhood has never sufficed to keep peace among them. When one reflects that even sons of the same parents, reared in the same home, can become bitter enemies when they disagree over the division of their inheritance, or when they embrace different creeds or ideologies, this negative result is not surprising. On the other hand, people of different races become close friends when they share ideals or loyalties. From this it is clear that human solidarity depends more upon having a common object of loyalty than upon the noble, but largely ineffectual, concept of the brotherhood of man.

Religions based upon unproved assumptions and irreconcilable dogmas have failed dismally to unite mankind. Conflicting political ideologies are a standing threat to man's existence. Two recent international organizations established primarily to preserve peace, the League of Nations and then the United Nations Organization, have proved deplorably incompetent to attain their major objective. International trade, nations' dependence on each other for raw materials and markets, is at best a fragile bond between them. Even the vast cost in wealth and human suffering is not a deterrent to war. It seems that there remains, as an effective bond between all peoples, only their common dependence upon the world that supports them and, thence arising, a common will to keep it flourishing and a common loyalty.

It will be evident to every thoughtful person that only by the united efforts of mankind can our planet's health and beauty be preserved. Nature knows nothing of man's often artificial and frequently shifting political boundaries. The winds blow over them as though they did not exist; great rivers flow past them without a pause; migrating birds fly across them without passport or visa; ocean currents are ignorant of the sovereignty of the shores that they wash; plants and animals are distributed over the world without the slightest regard for its political divisions. The planetary circulations are as international as the postal system, which so far has been man's most successful form of international cooperation. The "one world" of which we so often hear is the wider natural world and only hopefully the human world. On a planet which has been made intimately small by modern methods of transport and communication, the unity of nature demands the unity of men, whose existence depends on it.

Men may hate people who differ conspicuously from themselves in color, physiognomy, religion, or political ideology, but only the basest sort of man can hate the world that supports him. If the thought of the suffering inflicted upon the hated enemy cannot deter nations from war, perhaps the consideration of the injury wrought upon the unoffending planet itself might do so. Modern warfare inflicts such ugly and enduring scars upon Earth's face, causes such vast destruction of its nonhuman as well as its human inhabitants, squanders such immense quantities of its resources, that no one with even a rudimentary appreciation of what he owes to the natural world would dream of waging war. Certainly this is too fair and bountiful a planet to be ravaged by man's stupid quarrels. All men of goodwill might unite in loyalty to the world that bears them, if to nothing else.

Service to something greater than oneself brings a sense of personal worth and fulfillment that is too frequently lacking in contemporary life, and in such service we often find our greatest happiness. To be united in the endeavor to preserve the beauty, the fertility, and the living mantle of the planet that supports us all, along with the purity of the atmosphere that all of us breathe, would provide the firmest possible foundation for human solidarity, prosperity, and felicity. The recent widespread upsurge of concern for the preservation of the environment is one of the most heartening and promising movements that this world has seen for a long while.

Conservation is where ecology and ethics meet. To guide our efforts to keep land and water and their living communities in a flourishing state, we must look to the science that studies organisms in relation to their total environment, which includes all the other organisms that in any way influence them as well as all the nonliving features of their habitat, such as soil, water, and atmosphere. Since ecology is not only one of the newer sciences but also one of the most complex because of the vast diversity of its subject matter, we cannot expect it to attain the precision of physics or chemistry; we should be neither surprised nor discouraged if sometimes its recommendations lead us astray.

Although the practice of conservation involves ecology, its motivation and claim upon us fall clearly within the scope of ethics, which from ancient times has involved the examination of the good life and the attitudes, volitions, and conduct that conduce to it. The exhaustion of good land by vicious agricultural practices may, in the long run, cause more misery and death, by famine, than would a few outright murders. What could be more wicked and heedless of human welfare than to bring more and more children into an overcrowded world, a world so impoverished by ruthless exploitation that they hardly have a fair prospect of a satisfactory life?

Western ethics, from its foundation in ancient Greece down to modern times, reveals the pathetic blindness of philosophers who passed most of their lives in cities that were often walled and who gave little thought to man's dependence upon the natural world. They have commonly failed to see that the quality of our lives is affected by our treatment of land, water, and vegetation no less—and possibly more—than our treatment of each other, or how abuse of animals and even plants reacts upon character as disastrously as injustice to one's fellow men. Now, happily, concern for a deteriorating environment is expanding the scope of our morality, even though we may not recognize this. From another point of departure and with quite different principles, we are slowly approaching the breadth of moral vision revealed by the edicts of Asoka. Concern for

the environment may at last bring East and West closer together spiritually than religious ecumenism has succeeded
in doing.

As has been said, we owe our human form and innate
capacities to a long run of exceptionally good luck in evolutionary gambling. In a sense, the emergence of an animal
with such human attributes as rational thought, foresight,
compassion, and love of beauty and knowledge was not
merely man's good luck but a triumph for life itself, which,
after long experimenting and probing in many directions,
had at last risen so far above the struggle for bare survival
that it could begin to give free expression to higher qualities. The emergence of the human mind was not only man's
good fortune but that of Earth that bears him.

But was it? It has been bitterly remarked that nature went
mad when it created man, the most destructive of its multitudinous offspring, who threatens to annihilate the rest of
the living world and, finally, himself. Lately the rate of destruction has accelerated alarmingly, while there was never
before such widespread indignation over what is happening,
such general determination to mend our ways and live in
greater harmony with the natural world.

As evolution proceeds, the pace of change grows swifter.
The Precambrian era of the earliest, simplest forms of
aquatic life is estimated to have lasted 2,000 or 3,000 million years. The Paleozoic era, in which life emerged from
the seas and covered the land with luxuriant plants, among
which primitive insects and amphibians flourished, lasted
about 375 million years. The Mesozoic era, which saw the
genesis of flowering plants, birds, and mammals on continents dominated by huge reptiles, continued for about 160
million years. The Cenozoic or the Age of Mammals lasted
only about 63 million years. Man appeared in something
like his present form probably not more than 2 million
years ago. Only within the last ten or twelve thousand years
did he develop agriculture and, on this foundation, raise
cities and greatly increase his numbers.

Change continues at an ever more vertiginous pace. Considerably less than a century has passed since the first clumsy heavier-than-air machines made short flights, and now the skies everywhere are crossed by huge airplanes linking all the continents and inhabited islands. Time grows short. The next fifty years will probably tell whether man, unable to regulate his reproduction and moderate his excessive demands upon Earth's resources, will wreck the natural foundations of his life—and with them himself—or whether, awaking belatedly to a sense of responsibility for the planet that he dominates, he will become "the lord and not the tyrant of the earth" and usher in a new and glorious era of intelligent, compassionate stewardship of the exceptional planet on which it is his privilege to live.

I am not among those who believe that, in making man, nature exhausted its possibilities for creating a rational animal capable of developing a high culture, or that a major atomic war will destroy all the life on this planet. Undoubtedly, the fall and extinction of man, whether as the direct result of excessive crowding or in a conflict with nuclear weapons, will be accompanied by the appalling destruction of nature. But life is resilient, and much will survive on land and sea. If the biosphere is heavily contaminated by radioactive fallout, the rate of mutation and evolutionary change may be greatly accelerated. During the long ages that the Sun is expected to continue sending life-supporting rays to Earth, many changes will occur in its flora and fauna. Who knows what lemur or monkey, what small rodent or bird, what lizard or fish, relieved of domination by man, is destined to become the progenitor of a new race of rational animals that will dominate Earth and, we hope, rule it more wisely and benignly than man has hitherto done?

If we wish to continue to enjoy the privilege of living on an extraordinary planet, we can no longer rely on our luck. Evolutionary forces have ceased to act upon man as they did upon his remote ancestors; mutation continues, perhaps at an accelerated rate in a world contaminated by radioactivity; but such selection as occurs is unnatural rather than natural and is on the whole dysgenic, tending to lower the average quality of mankind, physically and mentally. Only a widespread determination to take rational control of our destiny, to stop our wasteful rivalries and quarrels, and to make a united effort to balance our account with the natural world that supports us can save mankind from a rapid descent of the downward path that has led countless other species to extinction.

14. The Expansive Spirit

 The ascent of consciousness is like the ascent of a mountain: the higher it rises, the wider its outlook, the more it strains its vision toward a far horizon. The human spirit has been called infinite, an exaggeration that points to a truth. The awakened spirit yearns toward infinity; it strains to reach infinity; perhaps only infinity, were it attainable, could quench this thirst and completely satisfy it. An outstanding attribute of an awakened spirit is its expansiveness, its insatiable hunger to experience more widely, to know more broadly and profoundly, to cultivate friendly intercourse with the whole of Being. The noblest mind is that which understands, appreciates, and loves the largest segment of the Universe. This yearning for the infinite, for the Whole, sometimes takes the form of striving for union with God, the ideal being who knows and embraces all.

The surrounding world both encourages and represses the spirit's innate expansiveness. By the sublimity of the heavens, the grandeur of land and ocean, the beauty of living things, the joyousness of young creatures, the generous and kindly deeds of men, it invites the spirit to expand and nurtures its capacity to appreciate ever more widely and intensely. On the other hand, the violence, harshness, and ugliness so widespread in the living world, together with the cruelty, selfishness, and perfidy of men, cause the spirit to contract, to draw away from the outside world and seek refuge within itself, like a mollusk tightly enclosed in its shell. Too wide an experience of the world's harsher and uglier aspects, too much suffering and too little joy, may cause us to hate rather than love the Whole of which we are parts.

From this loftiest aspect of the human spirit, we might derive a new and more penetrating conception of good and evil and a more comprehensive morality. Good is that which accords with the spirit's natural tendency, causing it to expand more widely and appreciate more intensely; evil is what opposes this natural tendency, causing the spirit to shrink within itself.

The most widespread form of the good is beauty, which, above all, invites the spirit to hold communion with the world and to love. Indeed, when we contemplate the immense, heterogeneous array of things that we call beautiful, we may conclude that the only property they have in common is their power to attract us, in a way that is often difficult to explain, and to win our love or admiration. Beauty sweetens our awareness that other things coexist with us and that we are not alone in the world. Although primarily applied to sensuous impressions, the adjective "beautiful" is quite properly extended to the conduct and even the invisible attitudes of people, when these are such that they make us proud to be human beings and to dwell among such fellow citizens.

The dull, uncultivated mind is interested in scarcely anything beyond the sensations of the body in which it dwells. As the spirit grows, it sprouts tendrils that fasten themselves to surrounding objects, living or lifeless, or to other minds. According to our temperament, the tendrils that we grow are either intellectual or aesthetic or a combination of the two. Some attach themselves to surrounding things chiefly by knowing about them; others, largely by bonds of feeling and sympathy. The more expansive our spirit, the more numerous are the tendrils we sprout, the more varied the objects to which they cling.

Youth is the proud and jubilant stage of life, because then these spiritual tendrils burgeon forth most easily and in the greatest numbers. We experience a sensation of growth, of expansion, that is rare at later stages of human life. Almost daily we can point to some new mental acquisition, to fresh spiritual growth. Then we are rarely oppressed, and never for long, by that feeling of the cessation of growth, of stagnation, which may cast so dreary a shadow over all too many of our later days. Yet, if we are wise, we will spare no effort to make spiritual growth continue through all the years of our lives. If the studies that have hitherto absorbed us fail to yield fresh insights—probably not because we have sounded them to the bottom, but because we have run out all our sounding line—then we should seek new ones. If old enthusiasms grow stale, we should try to cultivate others. Perhaps it is a wise practice to find a new interest every few years, so that we may never lose the sensation of growth and our minds may stay youthful.

The objects that stimulate the growth of our tendrils are as diverse as our temperaments. Music, minerals, sculpture, seashells, art, ants, the stars, orchids, antiquities, mosses, butterflies, birds—the list is long. The pursuits most to be preferred are those that promote the growth of tendrils compounded of both knowledge and feeling; they are strongest and most enduring when these two strands are interwoven. Whatever encourages these attachments—whether the stars that shine above us or the mosses that form a green carpet over the face of a rock—should be treated with the greatest affection and respect, even with reverence, for it invites the spirit to reach beyond itself and send its tendrils groping through the unfathomable mysteries of Being. Any study pursued with earnest enthusiasm, motivated by love rather than gain or fame, is capable of producing this effect.

To those who love nature in any of its myriad aspects, no tendrils are more sacred and dearly cherished than those which we attach to living things. A garden, wrote Francis Bacon, "is the purest of human pleasures. It is the greatest refreshment to the spirits of man." And, more than three centuries later, Donald Culross Peattie began his beautiful account of the lives and achievements of great naturalists with much the same sentiment: "Of all things under the sun that a man may love, the living world he loves most purely."[1]

The tendrils that attach us to the objects of our love and appreciation also form the strongest, most enduring bonds between mind and mind. Our business transactions are undertaken for gain; our profit may be another's loss. Our social intercourse is often governed by the desire to dazzle or to dominate, to be considered more clever than Mr. X or more charming than Mrs. Y. But, when two people who share the same enthusiasm for nature meet, their conversation soars higher. They tend to forget self and material gain, while they share their treasures of knowledge and experience. In giving no less than in receiving, the treasuries of their minds are enriched; old memories and discoveries of distant years acquire fresh life and color as they are recounted to a receptive mind.

These tendrils, bonds of knowledge, sympathy, and appreciation that unite us to what is beautiful, lovable, and true in the world around us, serve as a "flowery band to bind us to the earth." So long as they remain firmly attached, the expansive spirit is in no danger of shrinking into itself, violating its own nature.

As the most widespread form of the good is beauty, so the most common forms of evil are ugliness and harshness. The ugly thing repels us by its own nature, its peculiar appearance or organization, or perhaps more often by its lack of organization, the incongruity or disharmony between its several parts or aspects. One of the ugliest things in the world is the mangled corpse of any animal, for there, where we expect a high degree of coherence and symmetry, our shocked vision is greeted by incoherence and hideously distorted limbs and organs. However, our first reaction to certain living animals by no means lacking in symmetry

[1] Donald Culross Peattie, *Green Laurels* (Leipzig: Albatross, 1937).

and coherence is that they are ugly. This may be due in part to their strangeness, their wide divergence from the human form and that of the animals most closely associated with us; but frequently we consider them ugly because they menace us, actually or potentially, with fangs, stings, horns, or talons. We call them ugly because they are, or appear to be, harsh. By "harsh" I understand whatever injures or destroys any other creature, of its own or another kind. Cruelty is deliberate harshness, the infliction of pain or injury for the perverse pleasure that it gives a twisted mind. Just as we call generosity and compassion beautiful, so we recognize that harshness and cruelty are ugly. Beneath every experience of ugliness lurks, patent or concealed, resentment that the hideous thing shares Earth with us. If we cannot remove ugliness beyond sight and hearing, we try to flee from it.

The scope and relevance of our ethical judgments are greatly amplified when we recognize that the good is what stimulates the spirit to expand and appreciate, evil what makes it contract and condemn. Visible beauty, nobility of character, kindly deeds, at whatever distance from me in time and space, are to me, if I know about them, a positive benefit; although they may fail to affect me directly, they make my spirit expand in sympathy and strengthen its loyalty to the cosmos. Contrariwise, ugliness and harshness, wherever they occur, are harmful to me if I know about them, for they make my spirit shrink into itself and alienate it from the Universe. The sight of one animal striking down another in the wilderness would be considered by many to be morally irrelevant to me; nevertheless, it hurts me personally, for I resent living in a world where such acts occur, and I tend to contract away from it. Similarly, the harmful effects of one person's crime against another are by no means confined to the immediate victim and his family but extend in some measure to every morally sensitive person who happens to learn about it, for the presence of wicked people among us diminishes our respect and love for mankind and makes us less happy to be members of the human species. For the same reason, writers, whether of novels, plays, or philosophical essays, who for gain, or because of unhappy circumstances in their own lives, distort our perspective by magnifying the baser side of humanity and disparaging its virtues must be regarded as enemies of mankind.

Likewise, crudeness and discourtesy are positive evils, for they alienate us from our fellow men, whereas gentility and courtesy draw us closer to them. Poverty, truly so called, is a great evil, because it pinches people in body and spirit; incessantly preoccupied with finding the means of subsistence, the undernourished poor have neither inclination nor energy to cultivate the spirit and foster its natural tendency to expand in appreciative awareness of all things beautiful and lovable. Whatever discoveries or inventions ease the burden of living and free people to cultivate their higher faculties must be regarded as instrumental goods. Unfortunately, the excessive multiplication of laborsaving gadgets may oppress rather than liberate the spirit. Moreover, the opulence that comes from agricultural or industrial proficiency can be as spiritually devastating as poverty, for too often it encourages sensuality. The sensual mind tends to become imprisoned in its body, unappreciative of the delicate harmonies of the surrounding world. Its tendrils are few and short.

What chiefly alienates us from inorganic nature is its occasional violence, in the form of earthquakes, floods, hurricanes, and other catastrophes that unfeelingly destroy myriads of living things along with some of man's grandest creations. At greater distances, we learn of stellar explosions of inconceivable magnitude, which would vaporize any planets that revolve around these disintegrating stars; and astrophysicists tell us that our Sun may also, ages hence, burst like a titanic bomb, annihilating the planet Earth and all its life. The more we brood over such possible disasters, the more we shrink away from the physical Universe and recognize that it contains much evil. How can we feel solidarity with a Universe that appears so utterly indifferent to ourselves and all that we cherish?

Yet, if the inorganic Universe can destroy us, we must also acknowledge that it created us, for we are the natural outcome of processes that were active aeons before life arose and culminated in self-conscious beings, including ourselves. From day to day, the physical Universe, including the incandescent Sun and the rain-washed Earth, not only supports our lives but enriches them with the beauty of the starry sky, sunrises and sunsets, snowy mountain peaks, and the heaving oceans. And doubtless, if we could penetrate deeply into the hidden springs of the world process, we would discern that our presence here, with all our hopes and aspirations, is no accident but a partial fulfillment of a nisus or striving that stirred in the heart of the Universe from its prime foundation. Contemplation of the physical Universe reveals more that invites the spirit to expand than makes it contract, more good than evil.

Among the greatest of evils is senility. By diminishing our capacity to perceive and appreciate the things around us, senile decay inevitably makes the spirit contract. This painful deterioration is, in a measure, compensated by the greater understanding and sympathy that the years bring to a generously endowed mind. Memory substitutes for immediate experience; dim impressions from the environment may mean more to a mature mind than vivid impressions to a callow one. But in extreme senility the mind itself may so deteriorate that it loses contact with the world.

If all that stimulates the spirit to expand is good and all that forces its contraction evil, it follows that the culminating evil is death. If death is what it appears to be, the utter extinction of consciousness, it is the spirit's contraction to zero, its complete annihilation. We shall no more gaze upon the stars, nor see the faces of loved ones, nor hear the cheerful songs of birds, nor study the ways of the creatures that surround us, nor admire masterpieces of art, nor pore over the secrets of ancient rocks—all the sweet and tender bonds that we have long cultivated, all the tendrils that attach us to the good and the beautiful, will be severed. This sundering of cherished bonds, perhaps even more than the separation of spirit and body, makes death a gloomy and fearful thing to contemplate. If there be any evil greater than this, it is the spirit's expansion on the negative side of the zero point, in hatred and destructive rage, as sometimes happens to a sick soul.

Ancient philosophies and religions which taught that death, even if the total extinction of consciousness, is not evil reached this conclusion by concentrating upon the pains, sorrows, and frustrations from which it could release us, while they undervalued the rewarding experiences that life can bring. They regarded freedom from grief and anxiety as more desirable, or at least more attainable, than the spirit's expansion toward infinity. But, to those who recognize that the human spirit fulfills itself by expanding ever more widely in understanding and love, death is so great an evil as to be in a class by itself, a superevil. Only if it be, not annihilation, but a critical point in a continuing process of spiritual growth and expansion can death be regarded as other than a tremendous loss. If, released from the body, the spirit expands more widely than before, thereby fulfilling its natural tendency, death is no evil but a blessing in disguise.

While we remain in doubt about what lies beyond the dark gates of death, hardly anything can so mitigate the grim prospect of passing through them as the thought that, after we have gone, others who share our ideals will appreciate what we have appreciated, love what we have loved, care for what we have cared for, serve the causes that we can no longer advance. If we have given life to these others, or by our teaching or influence helped develop their minds and shape their ideals, and, above all, if in addition we love them personally, our departure will be less bitter to contemplate. If we cannot ourselves carry the torch that sym-

bolizes our aspirations to the final goal, we may, to para-
phrase a verse of Lucrétius, pass it still burning brightly to
others, like runners in a torch race.

We gaze into the starry night sky, and our spirits swell
with the vision. We stand beneath a noble forest, and our
spirits soar up to the treetops. We watch a limpid mountain
torrent sparkling in the sunshine, and our spirits dance
with the waves. We listen to the songs of birds and are glad-
dened. We contemplate beauty in any form, and our bur-
dens grow lighter. If we could explain these responses to
situations that contribute nothing material to our survival,
how they are related to the movement that made us and to
the Whole to which we belong, perhaps this enigmatic Universe would become less perplexing.

15. The Appreciative Mind

 When we reflect upon the vast stretches of time and the countless generations involved, we may concede that a process which depends upon random mutations and rigorous selection in the struggle for existence might shape the human body to its present efficiency, the mind to its actual sharpness, for these advantages promote survival. The strong passions that so often distress us are what we might expect in an animal that for ages had to confront a fiercely competitive world. But when we ask how a process that resembles a game of chance, with dreadful penalties for the losers, could have generated such qualities as love of beauty and truth, compassion, freedom, and, above all, the expansiveness of the human spirit, we are perplexed.

The more we ponder our spiritual resources, the more our wonder deepens. We can hardly avoid the conclusion that, stirring in the fecund depths of the Universe, an impulsion or urge that we can hardly conceive has been striving to lift Being to higher levels of awareness and value. Lacking omnipotence or foresight, this urge operates in the living world by the crude methods of organic evolution, blundering, taking unprofitable directions, but by ceaseless effort driving ever upward. The Universe, for all its immense age, might be compared to a youth with high aspirations, little experience, but indomitable persistence, who makes mistakes, suffers, but never ceases to try until he attains some of his goals. On this planet, we are the chief beneficiaries of this prolonged striving. The more he reflects upon the long ages, the immense effort, the struggle, and the suffering that were needed to make him what he is on a planet as richly endowed as Earth, the more the thoughtful, generous person appreciates and is grateful for his inestimable heritage. The gratefully appreciative mind stands at the apex of a long ascent from the first stirrings of sentience in living or, probably, lifeless matter.

We can hardly doubt that enjoyment is widely diffused over Earth. Despite occasional pains, sorrows, and frustrations, people enjoy their lives, as is evident from their reluctance to lose them. Animals, like children, seem to enjoy their food, their frolics, the companionship of others of their kind, their intervals of quiet repose. To believe that they do increases our estimate of the worth of Earth and makes us cherish it above such apparently lifeless planets as Venus, Jupiter, and all those more distant from the Sun. If plants find even the slightest satisfaction in their beneficent labor of photosynthesis in the sunshine, the value of our planet is greatly enhanced.

Although the presence of creatures that enjoy raises creation to a higher level and gives to Earth a significance that a lifeless planet could not have, this is not the highest which, even in our limited experience, it can attain. As we know all too well, people can enjoy without an appreciation of the sources of their pleasures; and such unappreciative enjoyment appears to be widespread among animals less thoughtful than ourselves. Appreciation adds to unreflective enjoyment grateful acknowledgment of the sources of our benefits and a sense of obligation to them. Enjoyment may be purely sensual; appreciation arises in thoughtful minds. The sensual animal, human or otherwise, cares only for the sensation; the appreciative person is concerned for whatever enhances his life. To the sensual animal, the sources of enjoyment are expendable; to the appreciative person, they are to be cherished and protected because they are intrinsically valuable. At whatever point in the evolution of life grateful appreciation may have arisen, it was a momentous advance for our planet.

Exploration of the contrasts between the enjoyment of

pleasures and an appreciation of their sources should help us understand the nature of appreciation. We may enjoy without appreciation or appreciate without enjoyment. Although many enjoy music, only a few are prepared fully to appreciate an accomplished performance. For this it appears necessary to have studied music enough to recognize the technical excellence of the composition and the competence of the musicians and to reflect upon the long years of training and practice that were needed to develop their skills. To have tried to write well increases one's appreciation of a polished style. To know the plants and animals around us enhances our appreciation no less than our enjoyment of a walk through fields or woods.

To appreciate what we do not enjoy requires finer qualities of mind than to enjoy what we are incapable of appreciating. An unwanted gift from one who desires greatly to please us may stir sincere appreciation without pleasure; and we may be grateful to the hostess who, to honor us, has thoughtfully prepared a dish that we do not relish. Appreciation may follow rather than accompany enjoyment; perhaps only in later years do we adequately appreciate all that was done by parents or guardians to make our childhood happy and fruitful. Or we may appreciate experiences that, far from being pleasant, were actually painful. In maturity, we may be grateful for the discipline and punishments, distressing at the time, that have helped us become self-controlled, responsible men and women. The mountaineer and the explorer, all who with great effort and hardship have achieved cherished goals, appreciate experiences that were painful and exhausting.

A most important difference between appreciation and simple enjoyment involves the treatment of their objects. The thoughtless pleasure-seeker is frequently careless of the things that gratify him. He wastes food, squanders or misuses Earth's bounty, mistreats painstakingly made artifacts, litters the ground with his trash. He may even impair his health and capacity to enjoy the sensations that he most craves. In contrast to the former, the appreciative person cares for whatever delights him. Grateful for the land's largess, he does not waste food or other resources. He cherishes the products of human industry and art. He leaves the spot where he picnics as lovely and unlittered as he found it or, perhaps, even cleaner. Likewise, he is careful not to diminish his capacity to enjoy or appreciate by overindulgence. He values things for their own sakes, not merely for the pleasant sensations they can yield. He husbands, never squanders.

It is, above all, in his relations with living creatures that the appreciative person differs profoundly from the unappreciative. The latter too often treats his dependents and those who serve him in any way as instruments or tools to administer to his comfort, pleasure, or profit. He ignores any qualities not useful to him and discards, like a worn-out garment, the person or animal who is no longer serviceable. To the appreciative mind, every animate creature is an end in itself, with attributes and capacities additional to those directly useful to him. He respects their individuality and never uses them as feelingless tools. Not only is he grateful for all services, including those that he has paid for, but, as far as he can, he helps his dependents develop whatever promising traits he might detect in them, even those of no direct value to himself. Those who pursue and kill free animals for pleasure appear incapable of appreciating their beauty, the mystery of the senses that guide them over vast distances, their devotion to nests and young, their unexplored capacity to enjoy and to suffer. One who appreciates the marvelous organization of a living animal could never needlessly deprive it of its life, least of all for the base excitement of killing.

It is evident that, although often closely associated, enjoyment and appreciation are different mental states, the first dependent largely upon sensations, the second upon more advanced psychic qualities. Without pleasure, joy, contentment, satisfaction with existence, and all similar affective states that we may include in the broad category of enjoyment, appreciation could hardly have arisen. Grateful appreciation is a precious addition to unreflective enjoyment, lifting it to higher levels. It appears to be a relatively recent development, and it may be increasing in the more thoughtful moiety of mankind. Indeed, unless appreciation, and all that it implies; grows apace, our sources of enjoyment will dwindle.

Like enjoyment, appreciation is often associated with love, yet it is not quite the same, and it will further clarify the nature of appreciation if we examine its relation to love. Before we do this, it seems necessary to explain the sense in which we use a word that is applied to the most diverse situations, ranging from sensual attachment to spiritual devotion. The outstanding characteristic of love appears to be desire to be near its object, although not necessarily in bodily contact with it. When it is absent, we yearn for its presence; when it is near, our happiness or contentment increases. Even the thought of being long separated from what we deeply love is distressing. Moreover, generous love not only cherishes and protects but strives to enhance its object, to increase its happiness or attainments if a person, to make it more perfect or enduring if a lifeless thing. In its highest expressions, love is simultaneously altruistic and egoistic, for by serving and enhancing what we adore we increase our happiness and self-esteem.

Although love and appreciation often go hand in hand, either may flourish with little of the other, for the former may consist of much feeling and little thought, whereas the latter may arise from much thought and little feeling. Accordingly, in childhood we tend to love more than we appreciate; but, as we grow older, we often appreciate much that we can hardly be said to love. If to be distressed when separated and comforted when reunited is admitted to be the distinguishing feature of love, a little child, even an infant, dearly loves his parents and home; but he can hardly appreciate their good qualities, or their efforts to keep him healthy and happy, or all that is involved in maintaining a well-ordered household. He may become so strongly attached to pets, dolls, or toy animals that he is inconsolable if they die or are lost, although after a few years he may detect little to admire or appreciate in things that he dearly loved. He lives in a fairyland where imagination and the affections predominate over the understanding of the sources of our satisfactions and delights that is the foundation of true appreciation.

As a generous mind matures, its love expands and deepens, but its appreciation spreads still more widely. We may appreciate the sterling qualities or good offices of people for whom we have a kindly regard that scarcely amounts to love, for we neither yearn for their presence nor are distressed by their continued absence. We may appreciate the benefits of a society that safeguards our freedom and produces what we need, yet prefer to dwell at a distance rather than in its midst. Although we may love one person, one home, or one animal with the deepest devotion, our love is often generalized instead of being centered upon particular objects or places. Ill at ease if not positively unhappy in a great city, one who loves wild nature yearns intensely to return to it, but many are the localities that can satisfy this longing. The bird lover is not so strongly attached to one individual bird, or even one species, that he cannot be happy almost anywhere that birds abound. Deprived of books, one who loves them feels a great void, yet his hunger for good reading may not be for one particular book or the works of one author. This expansion of interests and affections as the mind matures contrasts sharply with the condition of the child who is miserable when separated from the one home, the one mother, or even the one toy or pet that he loves.

The first requisite of appreciation appears to be associative thought. Our thoughts must pass beyond the sensation or situation that brings us pleasure or satisfaction to its source or sources; to the person who gives us a gift or performs a service for us; to the society in which we dwell contentedly; to nature which sustains and enriches our lives. The farther from our actual situation that our thoughts can reach, the more profound our appreciation is likely to be. Where other essential attributes of the appreciative mind are not lacking, knowledge and the studious habit greatly increase appreciation. Thus, one familiar with the history of man's prolonged, bitter struggle to win and preserve freedom of thought and other civil liberties will most appreciate the institutions of a free country. One who knows how long it took to develop high-yielding, nutritious plants and agricultural practices that insure continued abundance will most appreciate his daily bread. The biologist or paleontologist who has followed the immensely long, hazardous course of life's evolution is best prepared to appreciate his human body and all its high capacities. None better than the astronomer who has widely surveyed the Universe can appreciate the uniqueness of our planet and its generous hospitality to life.

Imagination that can give life and color to dry facts may heighten appreciation. To imagine what it must be like to live under a despotic government or in a police state should sharpen our appreciation of a free society. To imagine the life of a chronic invalid or an incurable cripple should increase our appreciation of unimpaired health. The mind that can visualize, even vaguely, the long ages that were needed to prepare our planet for advanced forms of life, and to create its many vegetable and animal species that serve or delight us, will most appreciate its present benignity. And, even more than imagination, actual experience of poverty, sickness, a depressing environment, or persecution will make us more intensely appreciative of present prosperity, health, pleasant surroundings, and freedom.

Spontaneous sympathy is an important element in appreciation. Often simple, poorly educated people, whose happiness so largely depends upon their relations with those around them, have more of this quality than more cultivated minds, whose thoughts range far from homely, every-day occurrences. These simple people may be more keenly appreciative of what others do for them, for little gifts and attentions, while those with greater amplitude of thought and knowledge are better able to appreciate what nature or human history has contributed to our welfare. Sympathy with the man or animal who toils and sweats in our service increases our appreciation of what he does for us.

Scarcely anything contributes more to appreciation than aesthetic sensibility. One who delights in the majesty of trees, the loveliness of flowers, the song of birds, the grandeur of landscapes, and all the beauty of nature can hardly avoid appreciating them and his life amid them. Responsiveness to art, including painting, sculpture, and music, at least in its more complex forms, appears to be less spontaneous and widespread, more dependent upon training and cultural influences, than responsiveness to natural beauty. Appreciation of art may be an acquired rather than an innate capacity, but one that can greatly enhance life, especially of those who live in great cities, far removed from nature. Unfortunately, aesthetic sensibility, one of the greatest spiritual assets of those fortunately situated, can become a severe liability. If forced to dwell amid ugliness and disorder, the aesthetically sensitive person suffers more than one whose sensibilities are duller, so that, instead of appreciating his existence, he may come to abhor it.

As haughtiness is a great enemy of appreciation and gratitude, so humility increases them. The person who believes that everything done for him, every gift, service, or distinction he receives, is no more than what is strictly due to one of his high rank or outstanding merit is unlikely to be appreciative or grateful. His attitude may resemble that of a despot of old, who believed that his subjects were created to serve his divine person, or that of certain citizens of modern welfare states, who are convinced that society owes them everything they cannot earn for themselves. The truly appreciative, grateful person is one who believes that he receives more than his due. This attitude of humble thankfulness may extend from life itself to the smallest gift or service; and, at whatever level it appears, it enhances appreciation and enriches life. We may reflect that we have done nothing to earn or deserve a well-endowed body and a perceptive mind but that they were bestowed on us as a free gift. At the other extreme, if we remember that the stranger of whom we ask a direction owes nothing to us, his courteous response will be more appreciated and will lighten our steps if the journey is long. If we never expect anything of anybody, we shall never be disappointed but will be more grateful for everything that is done for us. Even the return of a loan should make us thankful, for many a borrower has defaulted on his debt.

Although in moderation humility increases appreciation and sweetens life, like other virtues, it may be carried to an extreme that becomes a vice. To believe, as certain religious zealots have professed, that one is utterly worthless and deserving of nothing would destroy the appreciation of our splendid natural endowments that is the foundation of a truly religious attitude to life.

To be adequate and sincere, appreciation requires the generosity that recognizes the worth of those who help or serve us, the full value of what they give or do for us, and the excellence of everything in the natural world that supports or enriches our lives. We must not stint praise where it is due or fail to acknowledge merit where it occurs. To undervalue those who are kind to us, the worth of their services or gifts, or our delight in nature withers appreciation.

Among the most insidious enemies of appreciation is sensuality, which might be defined as excessive indulgence of the contact senses, taste and touch. Sensuality makes ends of means. It overindulges in eating and sex, which are means for life's perpetuation, not its ends. Accordingly, sensuality is akin to miserliness, which makes the possession of money an end rather than a means, and to fanaticism, which intensifies its means while losing sight of its ends.

Sensuality narrows the range of enjoyment and appreciation; by concentrating interest upon objects and situations that yield pleasure by contact, it distracts us from the far more numerous, varied, and elevating things accessible to eyes and ears. Although on the whole the contact senses tend to imprison the spirit in its body, whereas sight and hearing lead it outward, so great is our need to reach beyond ourselves that in the blind touch becomes more acute and compensates in some measure for the loss of vision—another example of the admirable adaptability of the human organism.

Acquisitiveness also narrows interests and limits appreciation. The number of things that we can possess is only a small fraction of those that we can enjoy without owning them. To admire without coveting greatly increases contentment. Instead of desiring the beautiful things that we see, we should be grateful that they exist for our enjoyment without our laborious care. This is eminently true of all the grand and lovely sights that nature presents to us and, even, of many of our friends' and neighbors' possessions— their gardens, the façades of their houses, the pictures on their walls. "The collector," wrote Anne Morrow Lindbergh, "walks with blinders on; he sees nothing but the prize . . . the acquisitive instinct is incompatible with true appreciation of beauty."

The appreciative person is restrained, even mildly ascetic, in his enjoyments. Glut of sensual pleasure and luxurious ease are foreign to his nature. He seeks experiences that expand his vision of the beauty and wonder of the larger world, rather than sensual delights that confine the mind to its body and may impair its capacity to perceive, understand, and gratefully appreciate. He hesitates to sacrifice to appetites or passions finely organized beings, with which he prefers to coexist harmoniously. He makes no unnecessary demands upon the productivity of a planet already overburdened with its excessive progeny. He lives simply and wholesomely, in order to remain, as long as he lives, a sensitive organ of appreciation, grateful for his blessings, diffusing love and goodwill.

The fine fruit of appreciation is gratitude. We may distinguish between appreciation of the gift and gratitude to the giver, but the distinction is somewhat forced. We appreciate the motives that prompted the gift, the giver's friendship or goodwill for us. When we turn to nature, and all that it contributes to our welfare and delight, it is more difficult to separate gratitude from appreciation. The traditionally devout may thank God for nature's bounty; but one who regards nature as a self-created and self-regulated system, dependent upon nothing beyond itself, may be profoundly grateful for its manifold gifts to body and spirit. Gratitude and appreciation are two aspects of the same mental state. Appreciation is commonly grateful appreciation.

The appreciative mind is concerned for the welfare of whatever it loves or admires; it cherishes and cares. Of caring we may recognize two degrees, *caring about* and *caring for* or taking care of. Our ability to care for is limited by our energy and resources and confined to the few things that we can nurture or protect, but no bounds can be set to what we care about. We care for our own persons, our dependents, our homes and gardens, and, if we live close to wild nature, the tiny segment of the natural world over which we exercise some control. One deeply concerned for mankind's future, for the beauty and fruitfulness of the unique planet that bears us, cares about things too big to respond to his unaided efforts. But, when enough people who care about great matters band together and make their voices heard, they may persuade nations to care for what they care about. Through caring, their grateful appreciation of nature's beauty may lead to effective measures for its protection. Through caring, appreciation of man's undeveloped potentialities may help produce a race more grateful for the privilege of living on so favored a planet and more responsible for its welfare. The most revealing aspect of our characters is what we care for and about, and this, in turn, is determined by what we appreciate. Our greatest claim to dignity, our most godlike attribute, is our ability to appreciate and care intensely for or about everything fair and good that Earth contains.

The capacity to appreciate, which so enriches our lives, is not without its darker side. As it watches helplessly the erosion and destruction of what it most values in the contemporary world, the appreciative mind can hardly avoid despondency. And even in the most favorable circumstances, as in a peaceful, prosperous world where men lived contently in harmony with nature, the mutability of all things under the Sun, the inevitable decay of what we most cherish, including our own powers, may induce spells of melancholy. Our greatest joys may be haunted by our greatest fears. As Keats perceived,

> in the very temple of Delight
> Veil'd Melancholy has her sovran shrine.

This brings us to the final outstanding attribute of the appreciative mind: it aspires. Not a world of misery and gloom, not a "vale of tears," is that best fitted to make us yearn for a happier existence. Such a world could hardly develop our most precious capacities or prepare us to imagine an excellent one. On the contrary, a world richly endowed, with much to enjoy and appreciate yet infected with strife, decay, and death, is most apt to foster our aspiration for life more prolonged, exempt from all the circumstances that distress us here.

Thoughtfulness, imagination, sympathy, aesthetic sensibility, humility, generosity, gratitude, freedom from sensuality and acquisitiveness, a great capacity to care for or about whatever it loves or admires, aspiration—these are the outstanding attributes of an appreciative mind. Although they depend in large measure upon innate qualities, they may be fostered by a judicious education. As these attributes grow, living becomes more rewarding and meaningful. Above all, the appreciative mind finds life significant. Boredom and vacuity are strangers to it, for, in a world so richly endowed as ours, it always finds something worthy of its attention, admiration, and devoted care. Born of passive contemplation, appreciation grows into satisfying activity.

An appreciative mind is an instrument on which the cosmos plays its tunes, a canvas on which the Universe paints its pictures. The whole course of cosmic evolution appears to have been directed toward forming such minds, which impart to Being its highest significance, and preparing a stage for their activities. By reverently and humbly acknowledging all that it owes to processes that long antedated its own existence, the appreciative mind becomes the holy or sacred mind. In simple natural piety, without theology or dogma, without making vast unprovable assumptions, it cultivates a truly religious attitude toward life and the cosmos to which it owes so much. Established upon immediate experience rather than upon ancient revelations or esoteric doctrines, its piety is unassailable by skeptic doubts. Grateful appreciation is the foundation of true religion.

16. Organs of Appreciation

By constantly shuffling the genes around in the most varied combinations, sexual reproduction has made each of us a unique being, an individual not quite like any other who ever lived. Differences in environment and education have added to our innate diversity. Most of us cling passionately to everything that expresses our individuality: our name and form, our opinions, our ideals and aspirations, our memories in which the continuity of our conscious lives is preserved, even those external possessions that most bear the imprint of our personality. Although there are, no doubt, certain aspects of ourselves that we desire to change for the better, there is much, including the most essential part, that we wish to preserve; to alter ourselves wholly would be to destroy our personality.

We do well to cling to our individuality. It is the product of an immensely long, difficult, and hazardous development. To create an animal with sensory organs like ours, capable of reporting so large a segment of the surrounding world to a mind that eagerly receives this information, classifies it, treasures it, and stubbornly endeavors to fathom its deeper meaning, was a triumph of the evolutionary process. A gratefully appreciative human mind is a center at which a substantial segment of the Universe is concentrated and valued. By the multiplication of such centers, the cosmos is immensely enriched.

A number of religions and philosophies have approved and encouraged this natural desire to preserve and perfect our individuality or selfhood. This is eminently true of Christianity, which promises to the righteous soul, purified of all defects and distressing memories, endless blissful existence in the sight of God, as beautifully portrayed in the concluding cantos of Dante's *Divine Comedy*. Jainism teaches that the individual soul or *jiva*, when cleansed of the karmic deposits that obscure its splendid innate powers, will preserve its self-identity forever, enjoying infinite knowledge, perfect bliss, and boundless power. Similarly, the Visistadvaita doctrine of the Hindu theologian, Ramanuja, affirms that the worthy purified soul will dwell everlastingly, as an individual, in God.

Among Western philosophers, Leibniz employed his great intellect to develop a world view that affirms the absolute indestructibility of the individual soul yet conforms to the laws of nature as understood in his day. We notice a remarkable similarity between the monads of the German philosopher and the *jivas* of Jainism. Both are indestructible; both contain within themselves, potentially, all that they can know. In the case of the *jivas*, this innate knowledge becomes explicit in the measure that the obscuring karmic dust, the product of wrong conduct, is dissipated; in the monads, the same result is attained by the illumination of darkness.

Although certain religions and philosophies affirm the value of selfhood and encourage the hope that it can be preserved everlastingly, others take an exactly opposite view. Not to cherish and perfect our individuality but to annul or destroy it, as by merging ourselves completely in the universal, the Absolute, or perhaps the void, is for them the proper aim of human life and the only road to liberation from the pains and tribulations that now afflict us. Notable among these religions and philosophies is the Advaita Vedanta of Samkara, which teaches that individuality is an illusion, for the *atman* or self of the apparent individual is no different from the *Atman* or Self of the Universe. The person who, through philosophic study or yogic discipline, succeeds in dispelling the mist of ignorance and recognizing his own true nature becomes inseparably one with the

Atman or absolute Brahman. He participates in the divine attributes of *sat*, *cit*, *ananda*—universal existence, consciousness, perfect bliss—but not as an individual, for his illusory self vanishes in the universal Self like a raindrop in the ocean. Similarly Buddhism, especially the southern Hinayana sect, is a discipline for the dissipation of the illusion of individuality, with all the suffering that appears to be inseparable from it. Aldous Huxley's erudite book, *The Perennial Philosophy*, advocates the submergence of self in the Divine Ground.

These religions and philosophies of regress, as they may be called, run counter to the whole course of evolution on this planet and, perhaps, in the Universe at large. Evolution, as Herbert Spencer demonstrated in great detail, is a progress from the diffuse to the definite, from the homogeneous to the heterogeneous, from the universal to the particular. Through this process arose, in the course of geologic ages, self-conscious individuals capable of knowing and appreciating the world in which they dwell—individuals whose selfhood is precious not only to themselves but likewise to others who surround them. By the presence of these individuals capable of living joyously, appreciatively, and responsibly, the value of the Universe seems to be vastly enhanced—indeed, without sentient beings capable of some measure of enjoyment, the Universe, for all its vastitude, would appear to have no value at all. But, say religions of the type of the Advaita Vedanta and Buddhism, this individuality is a tragic illusion, the source of all our woes, and we should bend all our efforts to annul it. For them, evolution, resulting in the genesis of individuals conscious of themselves and their distinctness from the rest of creation, was a blunder of cosmic magnitude, the evil consequences of which can be overcome only by the heroic efforts of an elect few. One who sees the situation in this light may well doubt that beyond the world of appearances lies a good or blessed Being, the source or eternal ground of this sad, illusory world, by union with whom (or which) we achieve salvation. For could a perfect, timeless Being make such a gigantic blunder?

Religious philosophies of the opposite type might be called religions of progress, because they try to carry forward to higher levels the main line of evolutionary advance on this planet up to the present era. Whether our individuality, or any part of it, does or can persist after the dissolution of our organic bodies is an open question. Objective science affords no proof of it. Yet the fact that in striving in this direction we are in the direct line of evolutionary advance should encourage us.

Which of these two types of religious doctrines, that which advocates progress toward an ever higher level of purified and disciplined individuality or that which advocates regress to the undifferentiated primal ground of Being, teaches the more valuable lesson? Which gives the truer picture of our status and function in a world where a vast amount of the good and beautiful is perversely intermixed with a distressing amount of the ugly and evil? Is it better to cherish and enhance our individuality or to strive with all our might to submerge it in the Absolute or the universal?

I believe that the doctrines of both types point to valuable truths but that either taken alone may develop a dangerously one-sided attitude, so that each needs to be tempered and complemented by the other. Our expanding scientific knowledge of the cosmos affirms with ever increasing authority the presence of the universal in our individual selves. Our bodies are composed of elements that spectroscopic analysis detects in distant stars. We live and act by means of the same energy that courses through the Universe, keeping the planets in motion, making the winds blow and bringing life-giving showers, enabling plants to grow, fishes to swim, and birds to fly. We belong to the great kingdom of living organisms, composed of cells and protoplasm like the rest of them, depending on the same Sun to support our vital activities, the widespread biological processes of metabolism, respiration, growth, and movement. To many, man's mental life appears unique; but the more attentively we study other animals, the more deeply in the animal kingdom we trace its roots. If we had instruments for the detection of spirit as sensitive as those that scientists use for measuring physical quantities, we might find that feeling is as widely diffused through the Universe as matter.

Although we are composed of universal elements, in us they acquire a unique configuration. All evolutionary advance consists in the arrangement of widespread, ancient elements in new patterns, more coherent in themselves, better adjusted to the surrounding world. In man, a recent arrival on this planet, the pattern of organization has attained an unprecedented level, endowing him with certain attributes unmatched by any other living thing. This union of the general and the particular gives to human life its outstanding significance. We can as little afford to lose sight of the one as of the other.

Our uniqueness and distinctness from the rest of creation, no less than our absolute dependence on universal forces and processes for the preservation of our individual lives, are stubborn facts that impress themselves the more strongly on our minds the more we contemplate them. Nevertheless, in consequence of faulty education or the influence of religious dogmas, we sometimes forget or depreciate one side or the other of our dual status, with lamentable results to ourselves and the beings that surround us. Forgetfulness of our universality and of the manifold bonds that join us not only to the whole living world but to the wider Universe confines our outlook, narrows our sympathy, and insulates our spirit from the cosmic currents that give to individuality its highest significance. Failure to appreciate and cultivate our individuality deprives our world of the most precious contribution that each of us can make to it: a unique center of awareness, knowing and responding to the cosmos in a distinctive manner and thereby contributing to its total diversity and perfection.

"The stage which most embraces the infinity of nature is the most individual," wrote the German philosopher Heinrich Steffens long ago.[1] To be an individual is not to set oneself in opposition to the Universe but to take the Universe, or as much of it as one can hold, into one's own mind.

The foundation of a fruitful spiritual life is the simultaneous awareness of our individuality and our universality, of our uniqueness no less than of our sameness with the whole of Being. This is a difficult position to preserve; it is fatally easy so to fix our gaze on either side of this duality that we lose sight of the other side. We need a symbol to help us keep firmly in mind our relationship to the Universe as a whole. The most adequate symbol that occurs to me is an organ of a living body. Such an organ—an eye, for example—is composed of the same substance as the rest of the body, nourished by the same food, dependent on the same vital processes for its maintenance. Detached from its body, it has neither life nor worth. Yet in it the cells of the body have acquired novel qualities and a unique configuration, which fit it for the performance of a special function, on which its peculiar contribution to the living body depends. The value of an eye would certainly not be increased if it became flesh and skin like the rest of the face in which it is set.

As an organ is to a living body, so is a human being, at his best, to the Universe. He is composed of the universal substance and kept alive by universal forces, which in him acquire a special configuration and operate in distinctive ways, thereby enabling him to make a unique contribution to the universal life. To view oneself as an organ of the Universe is to see in the most illuminating perspective our relation not only to the Whole but likewise to each of the other organs—the other individual beings—which surround us.

[1]Quoted in Harald Höffding, *A History of Modern Philosophy* (New York: Dover Publications, 1955), vol. 2, p. 199.

As organs of the Universe, what are our proper functions? Just as each organ of a healthy body exercises its peculiar power for the maintenance of the life on which its own life depends, so the person who views himself as an organ of the Universe will cooperate with his fellow organs in preserving the world in which he lives. That world maintenance is a human duty is by no means a new idea; it was fundamental to the Stoic philosophy and has nowhere been more beautifully expressed than in some fine passages in the *Meditations* of Marcus Aurelius. In the *Bhagavad-Gita* Krishna enjoined it upon Arjuna under the name *lokasamgraha*. As we see it today, world maintenance includes not only the cultivation of a healthy social order but likewise the preservation of Earth's beauty and productivity, upon which the prosperity of every society ultimately rests. The widespread forgetfulness of our organic relation to the natural world threatens to plunge it—and ourselves—into irretrievable disaster.

Although the primary or basic function of organs is to preserve the organism of which they are parts, they may also enhance the value of its life. This is especially true of such organs as eyes, ears, and other sensory receptors, without which our lives would hardly be worth living. One who regards himself as an organ of the Universe will wish not only to do whatever he can to preserve in a flourishing state that small part of it which he can influence but, moreover, to make some positive contribution to the value of the Whole. How can he do this?

A society watches the stars by means of its astronomers, follows the weather through its meteorologists, examines the structure of matter by means of its physicists, reconstructs its past by the agency of its archaeologists and historians. It is hardly an exaggeration to say that, when an astronomer peers through his telescope, the society that educated and supports him is watching the heavens; when a meteorologist constructs his weather charts, his society is mapping the state of the atmosphere; and so forth.

Our relation to the Universe, to the whole of nature, is more intimate and indissoluble than our relation to our society. We may abandon our society to join another or to live in solitude for years. We cannot by any means sever our bonds with universal nature. Accordingly, each one of us can, with no straining of meanings, view himself as an organ through which the Universe knows and appreciates itself. Through his eyes, it beholds the beauty that it has created; through his ears, it hears the melodies that it produces; through his nostrils, it smells the fragrance of its flowers; through his mind, it strives to understand itself and to fathom the mystery of its own existence; through his conscience, it passes moral judgments upon itself, approving what has been done well, condemning what is wrong. In him, if he is fortunate, it achieves the happiness or the satisfaction with existence for which, it seems, all Being hungers. It appears that only by means of individualized organs can the Universe bring to fruition potentialities that have long been latent in it, striving for ages to achieve realization.

In the Judeo-Christian tradition, the world was made for man. The Stoics held that the Universe existed for gods and men. The Universe was no more made for man than our bodies were made for our eyes. The Universe does not exist for man, but man exists for the Universe. Just as eyes, ears, and other sense organs evolved to enable animals to know their environment, so man evolved in the cosmic endeavor to know, appreciate, and understand the Universe. Made of the universal substance, formed and preserved from day to day by universal processes, we are to the Universe as our eyes and ears are to ourselves. If, in the long run, mankind fails to play its proper role as organs for knowing and appreciating the Universe, it will be discarded, as has happened in the past to many an unprofitable species that made enormous demands upon Earth's productivity.

Examination of our main sources of enjoyment reveals our fitness to serve as organs through which the Universe knows, appreciates, and seeks to understand itself. The sensations that most delight us, and bring the most enduring satisfaction, originate outside rather than within us. Conceivably, we might be so organized that our internal functions were our chief sources of joy. Every heartbeat might bring a thrill of pleasure; blood coursing through arteries and veins might arouse exquisite sensations; digestion might give us as much pleasure as eating delicious foods; every organ, from lungs and liver to the smallest gland, might yield pleasant feelings as it performs its proper function; muscles, instead of being chiefly a source of discomfort as they are forced to act when tired, or while they recover from fatigue, might with every contraction excite such a throb of joy that we would eagerly seek physical toil.

Although it would doubtless be pleasant to be so constituted that endogenous sensations were our chief sources of joy and sufficient for a wholly satisfying existence, we are otherwise made. Far from being delightfully aware of our internal organs and their operations, we hardly know that most of them exist unless we study anatomy, or until they are overworked or disordered, when they cause discomfort or pain rather than pleasant sensations. When in the fullest health, merely to feel oneself alive in a smoothly functioning body may be satisfying; but such satisfaction is chiefly experienced when we engage in some uncompelled activity that is externally directed, such as walking, swimming, or, as children, romping for the sheer joy of feeling ourselves alive. Of our wholly internal activities, the most pleasant, at least to contemplative minds, is mental, as in remembering, planning cherished projects, or thinking constructively; but the substance of our memories, fancies, or speculations has been provided by external experiences.

Our main sources of delight are not our internal organs but our outwardly directed sense organs. First and foremost, our eyes—such small parts of our bodies!—through which we view the beauties of nature and art and follow all the absorbing events of the surrounding world. Next come our ears, which register sweet melodies and welcome sounds or hear fascinating tales. Our noses inhale the fragrance of flowers, the aroma of fruits and other delicious foods. We eagerly seek these external sensations, without which our lives would be dull. Apparently, sight, hearing, and smelling evolved to their present acuteness because of their value in preserving the lives of hungry, vulnerable animals in a perilous world. Although, in an animal living in a state of nature, they are hardly overdeveloped for these purposes, they enable us to rise far above the exigencies of animal existence, to become appreciative spectators of the world around us, grateful for its beauty and sublimity, eager to know and understand it. Thereby we help fulfill the world process by realizing some of the high values that it can attain. Evolution blunders along, producing much that is horrible along with much that is splendid but, despite miscarriages, steadily raising creation to higher levels.

The more we explore the Universe, the more wonderful it appears to us. And perhaps the most wonderful thing about it is that certain infinitesimal parcels of its substance, organized in human form, reach out toward the stupendous Whole, trying to learn all that it contains, to discover its origin, fathom its meaning, and predict its destiny. Paradoxically, our greatness rests upon our recognition of our smallness and transience. To have a true conception of our smallness, we must be aware of the vastitude, immense age, and enormous diversity of the Universe. Although contemplation of the contrast between our dimensions, spatial and temporal, and those of the Universe may humble us, it should exalt us. The mind is as great as its thoughts. A creature who strives earnestly to know and understand so vast a Universe can never be insignificant, no matter how small its physical dimensions. Our eyes remind us that organs that contribute immensely to the value of existence can be small.

In the same sense that flowers are the goal of the plant, the end to which all its growth is directed, the fulfillment of its vital striving, so might appreciative, understanding organs be considered the ends or goals of the world process. We do not suppose that the plant plans its flowers or consciously strives to produce them. Nevertheless, its whole development is directed toward flowering, and, when it attains a certain degree of maturity or perfection, it bursts into bloom. Similarly, the world process is directed toward organs of appreciation, without which it fails to attain its highest significance. These organs are not necessarily humans; they include the whole company of intelligent beings, whatever their outward forms and whatever far planets they may inhabit, who respond to the wonder, majesty, and beauty of the cosmos with joyous, grateful appreciation, together with an effort to understand it and the determination to preserve, to the limit of their ability, everything excellent that it contains.

In addition to regarding ourselves as organs of the Universe, it is helpful to view ourselves as samples of the material, whatever our philosophy inclines to call it, of which the Universe is composed. The best sort of person is a revelation of hidden potentialities; he reveals the high degree of love, understanding, sympathy, and aspiration that the universal matter is capable of attaining, when properly organized. Is it not inspiring to reflect that the Universe contains vast quantities of whatever is needed to make the most admirable people who ever lived—of materials which, no doubt, are capable of forming beings more excellent than any that we know? To regard ourselves as samples of the universal matter should give us a new sense of dignity and importance, a fresh incentive to develop all our innate capacities and become more adequate revelations of the potentialities of the elements of which the Universe is composed. But, alas! the basest people disclose the depths of degradation to which the universal matter can fall.

By viewing ourselves in an organic relationship to the Whole of which we are parts, we steep our individuality in universality and give to our lives, which to some of our contemporaries seem petty and insignificant, a meaning and an importance that ennoble them. Moreover, we impart to the whole Universe, as far as it is known to us, a significance that certain recent thinkers have failed to detect in it. That which, by being known and appreciated, increases the value of a conscious life can never be held to exist vainly or barrenly. Even if, against all probability, we suppose that, apart from mankind, the Universe is totally devoid of purpose and significance, we must recognize that by responding to its sublimity and loving all the beauty that it contains we infuse significance into the Whole. Every person who, by simultaneously cultivating his individuality and his universality, makes himself a focal point in the cosmos enriches it by his presence. His universality gives meaning to his individual life; his individuality enhances the significance of the Universe.

17. Cosmic Loyalty

The mind that appreciates its blessings and strives to understand their origin becomes increasingly concerned about its relation to the Whole of which it is a part. Without a correct relation to the Whole, it is difficult to cultivate correct relations with the other parts. To orient ourselves to the Universe of which we are organs, it seems necessary to have a cosmology or world view. Such a view, with the attitudes and conduct that conform to it, seriously and devoutly held and practiced, constitutes the essential part of a religion or religious philosophy.

The great popular religions grew up long ago, when science was in its infancy and the prevailing accounts of the cosmos and the origin of life were utterly different from those that we hold today. These religions, especially those that arose in the Middle East and about the shores of the Mediterranean Sea, were products of a geocentric and anthropocentric view of the Universe that has been shattered by modern astronomy and biology. Although the ethics of these religions, born of the common experience of man rather than of cosmological speculations, still contains much of value, other parts of their doctrines are pitifully archaic. Attempts to reconcile them with our modern scientific view of the world are often childish or ludicrous. An earnest thinker cannot divide his mind into two unrelated compartments, one labeled "science" and the other "religion." To try to steer one's spiritual life by these antique beliefs might be compared to attempting to pilot an airplane with a sixteenth-century map of the world instead of modern charts. Some of us take our religion, our relation to the Whole to which we belong, too seriously to base it upon outworn concepts of the Universe. Not because we are frivolous or irreligious, but because we are too deeply religious, we cannot accept the popular religions.

With the notable exceptions of certain Eastern religions, the older religions begin at the wrong end. They start by affirming the existence of gods or a God, who are magnified humans, too often with the baser human passions, such as jealousy, vengefulness, anger, and greed; and from the supposed nature of their deity they derive their rituals and schemes of salvation. But for the existence of their gods or God they fail to present evidence that will withstand criticism. Every God conceived by popular religions is a lineal descendant of the gods created by the groping thought of early man, who could account for natural phenomena only by attributing them to humanlike beings. Or else the idea of God can be traced to the tendency to ascribe supernatural powers to deceased ancestors or rulers. Knowledge of how the idea of God originated is sufficient to weaken belief in such a being.

Although an anthropomorphic God might make his presence felt throughout an ancient kingdom, as its ruler did, we cannot imagine how a being that in any sense might be called a person extends his influence simultaneously through thousands of galaxies scattered through billions of light-years of space. The Universe disclosed by modern astronomy is far too big to be governed by any deity that we can conceive. And, if we cannot believe firmly in a God who accepts adulation, responds to prayers, rewards righteousness and punishes sins, how can we wholeheartedly accept a religion founded upon this belief? Moreover, if God exists, we cannot delimit him from the Universe; we cannot tell where this ends and God begins. If the world in its present imperfect state is an adequate expression of God's intentions, he is morally no better than the world. If he tries to make it more peaceful and harmonious but fails because of limited power, we have no way to know this. All that we can say of him is that he is a mystery beyond the mystery of the Universe.

Let us, then, try to develop a helpful or religious world view by starting at the other end—ourselves. If we have felt within ourselves a persistent longing for greater integrity and harmony, both in our personal lives and thoughts and in our relations with the beings of all kinds that surround us, and an unquenchable thirst for a good and happy life, we have an unshakable foundation for a helpful religion, for no one can make us doubt the reality of our own aspirations. When we trace this persistent yearning for a more perfect existence to its ultimate source, it becomes evident that it is a later phase of a movement or stirring widespread in the Universe, which massed the social atoms into stars and planets, brought forth life on Earth, and, with vast travail, raised life to its present level. Our longing for a more significant and harmonious existence appears to be an expression of an urge at the very roots of Being, become at last acutely conscious and foreseeing. Our grateful appreciation is our recognition of what we owe to cosmic processes. Here we have a link between what is most sacred and authoritative in ourselves and the Whole of which we are organs. We develop an essentially religious attitude and derive our ethics directly from an internal rather than an external source—all without building our whole ethico-religious structure upon unproved assumptions.

To view ourselves as organs of the Universe accords well with the unbroken continuity of the movement that stirred in the fertile depths of unorganized matter, created self-conscious minds, and finds expression in our hopes and aspirations. A healthy organ always functions for the benefit of the organism to which it belongs, even to the point of exhaustion. We might say that the organ is implicitly loyal to its body. Similarly, one who views himself as an organ of the Universe will be loyal to it, not unconsciously but with a patriot's devotion to his country or with even greater fervor, for we may change our nationality but never our absolute dependence upon the Whole of which we are parts. To be loyal to the Universe means, above all, to support with all our strength, intellectual, moral, and physical, its agelong striving to give greater value to existence. The most obvious and effective way that most of us can contribute to this effort is by doing all that we can to promote harmony, in our own lives, in society, and with all the living things that surround us, for harmony is the foundation of all higher values. Cosmic loyalty is devotion to harmonization.

An important aspect of loyalty is affiliation or solidarity with its object, whether this be a nation, a political party, or a cause. The loyal person does not hold himself aloof or deny his association with his country or cause. On the contrary, he is proud to affirm his close connection with it. Similarly, one who cultivates cosmic loyalty does not separate himself, mentally, from nature, which is the cosmos viewed under another aspect, the totality of its ongoing processes. To view man as distinct from nature and to compare the two to the detriment of the latter, as in our overweening human pride we too frequently do, is not only unfair to nature, depriving it of what rightfully belongs to it, but it hurts us by alienating us from our origin and the source of our strength. Too often we hear it said that man has a conscience but nature has not, that man is (sometimes) compassionate but nature is cruel, that man is foreseeing but nature is blind, and similar contrasts.

To disparage nature in the foregoing fashion is logically equivalent to saying that our eyes see but our bodies are blind, our ears hear but our bodies are deaf. To be sure, if we gouge out our eyes or rupture our eardrums, our bodies will be blind or deaf, but only because we have deprived them of what rightfully belongs to them. We, at our best, are the organs that nature at long last has developed for the expression of conscience, compassion, foresight, and similar mental or spiritual qualities. To deny that they belong to nature because we cannot detect them throughout the natural world is just as perverse as to deny that vision is an attribute of the human body because the body does not see with its whole surface. If we refuse to recognize that man's compassion, or any other quality of which we are proud, belongs to nature, to what shall we ascribe man's fierceness, which at times exceeds that of any other animal? The tiger's ferocity and human compassion are equally revelations of the potentialities of the cosmos; and, as later developments, compassion, conscience, aspiration, and the like appear to be truer indications of the direction in which evolution has been proceeding and, doubtless, will continue to proceed, if we loyally cooperate, using the great endowments that nature has already given us. To remove man, in thought, from nature impoverishes our concept of nature without raising our estimate of man.

The price of the disloyalty that refuses to acknowledge that man belongs to nature is alienation, a chilling sense of being alone in the Universe that can become despair. Such disloyalty severs us from the source of our strength, nature's boundless creativity, which accomplishes so much that we understand imperfectly or not at all. When we consider how the scarcely visible speck of organized matter that is a fertilized ovum can direct the development of a large animal containing hundreds of organs and billions of cells, how without chart or instruments birds migrate back and forth between two definite spots thousands of miles apart, or how insects respond to the most tenuous signals, it would be absurd to set arbitrary limits to what, as members of the natural community, humans can achieve—not only in technical feats, such as placing men upon the moon, but in intellectual and spiritual elevation. Let us humbly and gratefully acknowledge the source of our powers, clarify and purify our ideals, and proceed unitedly to learn what nature will do for us if we cooperate intelligently with it.

Cosmic loyalty no more implies approval of everything that happens in nature than patriotism means approving all that happens in one's country. The truest patriot is one who condemns most unequivocally every abuse, injustice, and shortsighted policy that he detects in his nation; to close one's eyes to wrongs that might be righted is a sort of tacit treason. Similarly, to proclaim that all that is must be right is treasonable to nature or the cosmos. If we concede that we are parts of nature, then we must regard our abhorrence of all the carnage and suffering that we behold as nature's protest, from a higher level, against crudities that were apparently inseparable from life's onward march. Even if we can do no more than feel and express indignation, our strong disapproval marks a momentous evolutionary advance that is perhaps a promise of greater things to come. In a world where good and evil are so intermingled as in ours, to be organs of protest against all that is harsh and ugly is no less necessary than to be organs of appreciation of everything beautiful and excellent. In both ways, we show our cosmic loyalty.

Just as loyalty to our country does not save us from being oppressed by stupid or unjust laws, occasional violence from fellow citizens, or unfair accusations in the courts of law, so loyalty to the cosmos does not guarantee us immunity from nature's occasional rampages. We are vulnerable to volcanic eruptions, earthquakes, tornadoes, floods, and other manifestations of cosmic forces, which do not temper their might to spare the most loyal heart. Although in many ways we are the most favored of nature's children, this does not protect us from others of its teeming progeny, which may ravage our fields, consume the timbers of our dwellings, suck our blood, or invade our bodies with distressing or fatal consequences. Although nature has given us the ability to protect ourselves better than any other animal can, cosmic loyalty does not dispel all misgivings or make us feel absolutely secure.

Nevertheless, the greatest threat to man has from remote ages been man himself, not only hostile tribes or nations but also wicked individuals within our own society. Never have people been in greater peril from other people than today, when the world's most powerful nations have huge stocks of the most destructive weapons, and crimes of all sorts become yearly more frequent. If cosmic loyalty may be repudiated because nature sometimes deals harshly with us, a similar objection can be raised, with greater justice, against humanism, which proposes mankind as the object of religious devotion. Since there is much truth in the old dictum that fear made the gods, it is hardly an exaggeration to say that theistic religions were born of fear, and they certainly have not dispelled it. On the contrary, some of them, by filling credulous minds with dread of relentless punishment by an arbitrary, vindictive God, have intensified fear. If a loyal heart seeks an object of allegiance from which he has nothing to fear, let him be loyal to his own high ideals. He will never suffer violence from them.

To understand the immense difficulties of creation should fortify our cosmic loyalty. We do not know whether the Universe is eternal or of limited, although inconceivably long, duration. Astronomers have not agreed whether it is cyclic, alternately expanding to vast limits and contracting to an explosive "cosmic egg" in which all its matter is concentrated; or whether it is in a steady state, with fresh matter springing up to form new galaxies in the interspaces left by the everlasting recession of older galaxies. Conceivably, in past cosmic eras or in distant regions of the contemporary Universe, other planets have evolved beyond the point that ours has attained and might serve as models for ours to follow. But, even if such planets exist or existed, it is highly doubtful, in view of their immense distance, that any of them could serve as a model for our own.

With neither model nor guide, without foreseeing intelligence, the social atoms that compose our planet, starting from a low level of organization, have had to build up complex, coherent patterns by the slow, painful process of trial and error. No wonder that they have needed a long age to cover Earth with an immense variety of organisms that are still far from composing a harmonious living community. No wonder that they took wrong directions, creating whole families and orders of organisms that were eventually exterminated. That they have not made us perfect beings in a perfect world is only too evident. Nevertheless, our imperfections, physical and psychic, are overshadowed by our superb endowments, such as no other creature in the whole animal kingdom has received. Only an ignoble mind would be ungrateful for our endowments because of our defects, especially when, wisely used, the former can in numerous instances remedy or overcome the latter, as by making spectacles to correct poor eyesight or exercising self-control to subdue passion. What excites our admiration is the indomitable persistence of the creative energy which, despite all obstacles, has produced so much whose excellence commands our approval. Remembering the lack of guidance, the immense problems to be solved, the generous spirit will forgive the creative movement its crudities and blunders and be profoundly thankful for its achievements.

The accomplishments of the social atoms, continually rearranging themselves in patterns of ever increasing complexity, coherence, and amplitude, appear the more wonderful when we compare their achievements to major human inventions. These were made by intelligent men who had definite goals and were familiar with the mechanical or electrical principles that a long line of predecessors had developed. Nevertheless, they proceeded mostly by trial and error; their earliest models were usually imperfect or worthless; and it required a long series of improvements, by a succession of inventors, to bring their inventions to their present level of efficiency. Not surprisingly, an unguided process took a much longer time, and a much greater number of slightly improved models, to produce organisms vastly more complex and more closely integrated than any of man's inventions—moreover, organisms capable of self-regulation and self-repair to a degree that none of our mechanisms is. Most remarkable of all, the guiding intelligence, which might have been so helpful at the beginning, is at last emerging from this agelong process.

Cosmic loyalty is, above all, the determination to dedicate this intelligence, the fruit of such immense travail, to the advancement of the cosmic striving to give ever higher significance or value to existence. Even in the absence of explicit loyalty, intelligence has already done much, as in the arts both practical and fine, to enhance life—not, however, without involving humanity in serious difficulties, as by overcrowding and overexploiting the environment and multiplying injurious or highly dangerous productions. To extricate man from these difficulties may require firmer determination and more intelligence than were needed to get him into them.

To help people become more understanding, appreciative, loving, and loyal is one of the most urgent tasks of cosmic loyalty. Much can be accomplished by education, but without continuing evolution man is unlikely to rise far above his present level. Evolution requires selection which, in the virtual absence of the usual agents of natural selection, must be widely practiced by man himself if any significant, sustained advance is to be made. Although the conviction that, by careful selection, man can improve himself as he has improved his domesticated animals and plants is very old, social prejudices, lack of wisdom, and the absence of a clear, widely held ideal of what we should become still make eugenics difficult to apply.

Eugenics may be aristogenic or depurative. The aim of the former is to produce individuals who excel in certain attributes that are held to be most desirable, such as intelligence, courage, physical stamina and endurance, or whatever qualities are most esteemed by the eugenist. It encourages the reproduction of individuals who manifest the preferred attributes to a high degree, as in Plato's *Republic*. By segregating outstanding individuals in an elite group, it may drain the general population of whatever genes support the desired quality, thereby impoverishing the multitude. Depurative eugenics, on the contrary, tries to raise the quality of a population by discouraging the reproduction of its physically or morally least desirable members. Most obviously, it could improve health and happiness by preventing the multiplication of defective genes, such as those that cause hemophilia, muscular dystrophy, color blindness, and many others. Although these injurious genes continue to arise infrequently by mutation, their presence can be detected and their transmission to descendants prevented. Aside from this, we should wish to decrease the number of the stupid, the irresponsible, the violent, and those deficient in love and kindness. Much would be accomplished if we could prevent the birth of all babies, except those who are wanted and lovingly welcomed by physically and mentally sound couples who have made the necessary preparations and generously desire to give others what they themselves appreciate so profoundly: the experience of living in a well-endowed human body on a beautiful and interesting planet.

In recent years we have heard much about the prospect of improving the quality and productiveness of domestic plants and animals, and of man himself, by "genetic engineering." Instead of waiting for the appearance of naturally occurring mutations, as breeders of animals and plants have done in the past, or even increasing the frequency of random mutations by shortwave radiation or chemical treatment, the genetic engineer directly manipulates and rearranges the genes themselves by methods recently developed. This could, theoretically, be applied to implement a program of aristogenic eugenics. As in any such program, we would first need to agree upon which "noble" or useful attributes we wish to promote in the human stock, and past experience has shown that such agreement is difficult to achieve.

I surmise that genetic engineering would be more successful in man, as in other organisms, in altering such physical attributes as size, strength, resistance to disease, or longevity than in improving the moral and spiritual level of the race. It might be exceedingly difficult to identify and locate, much less improve, the constellation of genes that makes us generous, responsible, appreciative, and the like—qualities that appear to depend at least as much upon early environment and education as upon the genes themselves. Not to be overlooked is the tremendous cost of evolutionary change, which involves not only the appearance of new and perhaps "superior" genes but the elimination of the old, "inferior" genes that affect the same characteristics—a process so costly in lives that evolutionists have wondered how a population could survive it. We should be wary of entrusting mankind's future to manipulators of genes, who might readily become autocrats of unprecedented power. Random mutations through the ages have given mankind a valuable diversity of temperaments and talents that it would be a pity to lose at the hands of genetic engineers. Our safest course appears to be to try to reduce or eliminate the undesirable aspects of this diversity by means of depurative eugenics.

Often a loyal citizen can do most for his country by giving his home town or county the benefit of his energy. Similarly, we can best demonstrate our cosmic loyalty by serving our home planet—we can still do little beyond it. Possibly the time may come when man, grown immeasurably in wisdom and love, can raise the whole life of this planet to unimagined heights; but such a time will never come if Earth continues to deteriorate as an abode for life at the present alarming rate. The growing movement for the conservation of nature is, as we saw in chapter 13, inspired by diverse motives. Those whose motivation is most disinterested wish to protect the natural world because they value the excellent things that took so long to create; they are deeply distressed by the extinction of beautiful species or the spoliation of scenery that they have no hope of ever seeing. These conservationists are inspired by cosmic loyalty, although they might not call it by that name.

The attitude denoted by cosmic loyalty is by no means new in the world. Probably no creed fostered it more earnestly than Stoicism, which taught that the cosmos was governed by a benign providence, which ordered all things for the best. To maintain this position, it developed paradoxes, such as that pain is not evil, which their philosophical contemporaries ridiculed and which would hardly be accepted by modern people. The Stoics cultivated a certain emotional coldness which was, however, mitigated as their philosophy spread widely through the Greco-Roman world. Their stubborn loyalty to the cosmos is inspiring. Eighteen centuries ago, the Stoic Emperor Marcus Aurelius Antoninus wrote in his intimate journal:

> In the morning when thou risest unwillingly, let this thought be present—I am rising to the work of a human being. Why then am I dissatisfied if I am going to do the things for which I exist and for which I was brought into this world? Or have I been made for this, to lie in the bed-clothes and keep myself warm?—But this is more pleasant.—Dost thou exist then to take thy pleasure, and not at all for action and exertion? Dost thou not see the little plants, the little birds, the ants, the spiders, the bees working together to put in order their several parts of the universe? And art thou unwilling to do the work of a human being, and dost thou not make haste to do that which is according to thy nature?[1]

Here we have a thought that could be greatly expanded by a modern ecologist: the cooperation of creatures the most diverse to keep the Universe, or our particular corner of it, in a flourishing state. How much saner and more helpful is this than the idea of man dominating nature, putting all things under his feet, in the words of the ancient psalmist! Unless the governors of nations and private citizens alike can rise to

---

[1]Marcus Aurelius Antoninus, *Meditations*, trans. George Long, V, 1.

the grandeur of the philosophic emperor's conception of working together with the rest of the living world to preserve Earth's health, the prospect for our planet is bleak.

Man is the chief beneficiary of a process which, for billions of years, has been covering Earth with life and beauty, making it an ever more favorable abode for beings increasingly able to enjoy their existence upon it. No other animal is so well equipped to appreciate the planet's grandeur and beauty or to understand it. Probably in no other era has Earth had more to offer, more accessibly, to an appreciative or inquiring mind—at least, until the last few decades of unprecedented, worldwide devastation. Ought we not to be profoundly grateful for our status as the chief beneficiaries of a process continued for such immense ages, on so grand a scale, with such great effort? To be sure, the process has had miscarriages, from which we, and all creatures, suffer. But our benefits far outweigh our afflictions, and this should encourage us to use all our strength and intelligence to make our planet an ever more auspicious home for the creature that it has endowed beyond all others and for all those others—the great majority of existing species—that are capable of coexisting peaceably with him.

The success of a process directed toward increasing the value of existence depends not only upon what it can create but equally upon what it can preserve. The value of a beautiful object is greatly increased by its permanence, which permits more appreciative spectators to enjoy it for more hours. Nature keeps Earth lovely by constantly replacing its adornments rather than by preserving each of them indefinitely. Sunrises and sunsets fade, to be succeeded by others that are never exactly the same. Flowers wither, but more will bloom tomorrow. The loveliest and most amiable animals die, but their progeny inherit their charm. Whole species become extinct, but new forms evolve to fill the gaps. Sublime peaks are eroded to unimpressive hills, while others are slowly uplifted. Thus, by means of endless substitutions, nature keeps Earth green, fruitful, and splendid. Nevertheless, in the far-distant future, all life will be extinguished on this planet by the cooling of the Sun, if not more swiftly annihilated by the Sun's explosion. Will nothing less durable than atoms be preserved?

Nothing seems more worthy of indefinite preservation than loving, appreciative minds. Such minds, products of diverse heredity and experience, tend to be unique and irreplaceable to a degree rarely or never exhibited by individuals of species other than man. In the measure that they have found things worthy of appreciation and love, they cling to their existence and shrink from the prospect of eventual extinction, although they may school themselves to face it bravely. Their memories treasure, more or less vividly, precious experiences that may never be repeated, lovely things that have passed away, all of which would fade into oblivion with the extinction of consciousness. An alert mind, which for years has eagerly absorbed images from the surrounding world and tried to understand it, becomes a microcosm that mirrors, although imperfectly, the macrocosm. By the increase of such minds, creation is multiplied; by their extinction, it is diminished.

The indefinitely prolonged preservation of appreciative minds would be in the direct line of advance of a creative process that has steadily increased the value of existence; failure to preserve them would leave this process incomplete. Therefore, it is rational to believe that, if spiritual survival is possible, it has been, or will be, achieved. The relation of consciousness to the brain is so obscure that we do not know whether the two can be separated. We do know that a great variety of things, including some undetectable by our senses and probably also by our growing array of sensitive instruments, exist in space that we call empty.

Consider a small volume of space above a wide landscape on a sunny day. Continuously coursing through it are light waves from each of the immense number of visible points within the horizon. Add to these light from each of the myriad stars that the most powerful telescope could reveal, even the brighter of them invisible now only because the sunshine is so strong. Add to these the infrared waves from many points in the landscape and ultraviolet waves from the Sun. Add to these radio waves, from every station that a sensitive receiver might pick up and from distant galaxies. Add to these a gravitational field and a magnetic field, and I know not how many other radiations and forces more obscure. A volume of "empty" space much smaller than a human brain appears able to contain, in a single moment, a complexity comparable to that of the brain, for all its billions of neurons.

Not the least wonderful is the fact that such an immense throng of waves and forces can cross each other simultaneously, at a single point, without losing their identity, as though at a busy intersection continuous streams of speeding cars from a dozen highways could pass each other without ever colliding. We cannot reject the possibility of a disembodied spirit, with many thoughts and many memories, just on the ground that it would lack support for such varied contents. Space, the mysterious, the unexplored, the persistently misunderstood, can support infinite complexity in a nutshell. The chief problem is, can our consciousness be detached from our perishable brains?

Perhaps survival is, or will be, dependent upon the mind's quality. A sensual mind, engrossed in its own body, with little feeling for the beauty and sublimity of nature or for the finer human relations, with no high aspirations, might be lost and unhappy if detached from its flesh. An appreciative mind, filled with love and cherished memories of beautiful and interesting things, might continue to exist happily apart from the body that nourished its growth. Perhaps the mystics and ascetics who tried to win everlasting bliss by averting their eyes from the visible world and detaching their affections from its creatures were on the wrong track. Not the insulated or introverted mind but the outgoing mind, richly stored with what it has appreciated and loved, is more likely to have the power to exist without the support of a body. Conceivably, such minds might preserve memories of the planet they have loved and protected, long after Earth itself has ceased to exist as an abode of life— which would be the triumph of cosmic loyalty.

To aspire to everlasting life is a convincing expression of cosmic loyalty. Those who do not intensely desire to prolong their conscious existence beyond this brief span of organic life seem not to love enough, to appreciate enough the glories of nature, to thirst enough for deeper understanding, to value enough the high privilege of existing as a self-conscious being, to treasure enough the memories they have gathered, the vision of the cosmos that they have been forming. To view with indifference the extinction of our minds, with all their hopes, aspirations, memories, is certainly to value these things lightly. And this ardent desire to prolong our conscious existence comes to us from the cosmos itself: to perpetuate itself is the first principle of Being, of which we are parts.

Epilogue

The tropical rain forest is the maximum expression of nature's creative power. Massive trunks soar far upward to support the spreading crowns of an immense variety of trees, which at intervals cover themselves with blossoms. Scattered among the branching trees, slender palms rise tall and straight to uphold rosettes of great feathery or fanlike fronds. Twining about the trunks or hanging from lofty boughs like thick cables, lianas compete with the trees to expose foliage and flowers to the sunshine. Ancient trees support aerial gardens of orchids, bromeliads, aroids, ferns, mosses, and other epiphytic growths, including smaller trees. In the windless aisles between great trunks, stately tree ferns bear symmetrical crowns of wide, intricately patterned fronds. On the ground, smaller ferns flourish in subdued light, amid flowering shrubs and herbs.

Through this exuberant vegetation flit birds in great variety, the more colorful in the sunlit treetops, the more plainly attired in the shady undergrowth. From time to time their voices ring out, startlingly loud, calmly sweet, plaintive, shrill, or droll. In the day's middle hours, morpho butterflies float through the forest glades, flashing intense azure whenever a stray sunbeam strikes their wide, satiny wings. The more intently one searches, the more beauty he discovers; the more patiently he watches, the more fascinating patterns of life he discloses. An attentive lifetime is not long enough to reveal all the wonders of a hundred acres of tropical rain forest.

It is so easy to become absorbed in the immense diversity of sylvan life that one forgets its perils and tragedies. Snakes lurk amid the ground litter, quick to sink hollow, venom-filled fangs deep into the legs of the unwary. Although in tropical American forests they are rarely seen, to ignore them is foolhardy. Occasionally one comes upon the pitifully mangled remains of a furry animal, victim of a raptor or mammalian predator; or feathers forlornly scattered reveal the spot where a dove or some other bird has been struck down. Although rarely so troublesome as in some northern forests and thickets, mosquitoes and redbugs remind us that at all levels life preys upon or parasitizes life. To hatch their eggs and rear their nestlings in this predator-ridden environment, birds conceal their nests most cunningly or hang them inaccessibly. Despite these precautions, only one in four or five escapes predation until the young fledge.

Although less spectacular, the struggle for existence among the plants is no less intense. Of a thousand seedling trees, scarcely one will find an opening, left by the fall of a forest giant, where it can vie with many others for a place in the sunlight; most languish and die in deep shade. Even large trees are not immune from attack, for they may be killed by the coils of thick lianas that spiral upward around their trunks or by strangler figs which, germinating on high branches where birds have left the seeds, cast networks of slowly thickening roots around their boles. If it were concluded in minutes instead of years, the conflict between a fig tree or a liana and its arboreal victim would be as violent as that between a boa constrictor and its mammalian prey.

How one reacts to these contrasting aspects of the forest depends largely upon his temperament and interests. He may stand in reverent awe amid this supreme display of life's diversity, while towering trees draw his eyes upward and his thoughts follow, questing for the meaning of so much vital activity. If imaginative sympathy enables him to live in the flesh of other creatures, he may be saddened by the outrage of predation, oppressed by the intensity of the struggle to remain alive, dejected by the thought that, amid such profuse life, so much must be prematurely blotted out and wasted. Or he may alternate between these moods, much of the time so absorbed in tracing the countless complexities of tropical nature that he only occasionally meditates upon its wonder and mystery and even more seldom, on the rare occasions when he unexpectedly witnesses an act of predation, protests in anguished silence against such callous harshness.

This tropical rain forest, this maximum expression of terrestrial life, might be taken as an epitome of the whole living world. Throughout this world, we find the same contrasts of beauty and horror, peaceful growth and violent or lingering death. In no part of this world are the contrasts more glaring than in its human segment, where man's restless mind has intensified every natural trend, so that he not only nurtures more tenderly and constructs more grandly but maims and destroys with more frenzied rage. And, just as in the forest, one's reactions to this situation depend upon his temperament, education, and experience, with the difference that, touching us more intimately, the varying aspects of man may arouse stronger emotions. Some fortunately circumstanced individuals seem almost to dance through life, unperturbed by the misery of a large fraction of mankind. Others are worried or depressed by the human situation. Or they may vacillate between optimism and despair, depending perhaps on the state of their health or the latest demographic report. Some sensitive souls, well born, well educated, lacking nothing, are oppressed by a feeling of guilt that they have so much while so many hunger, suffer, and die. In extreme cases, they voluntarily renounce a comfortable life to endure hardships while serving the unfortunate.

This raises an insistent moral question. Is it wrong, or at least callous, to be lighthearted and happy in a world that harbors so much misery? Should we not, instead of regarding ourselves as primarily organs of appreciation of everything joyous, excellent, or lovely that the world process has brought forth, become above all organs of protest against or indignation toward all the suffering and ugliness it has produced? Instead of contemplating nature's beauty, would it not be nobler to plunge into the midst of man's ugliness and foulness and do whatever we can to cleanse and sweeten it?

To brood despondently upon all the suffering among mankind or in the whole living realm, calling it a vale of tears, helps make it just that. If no one could be joyous or happy so long as any creature is miserable, this would become an unbearably gloomy world. To suffer sympathetically because others suffer, even at a great distance, reveals admirable fellow feeling, but it uselessly increases the sum of unhappiness. Although a saint like Avalokiteshvara might defer the attainment of bliss until the least creature won release from pain, if each of us refused to be happy because at all times some are inevitably sorrowful, we would so depress one another that life would become unendurable.

In this predicament, it seems best to follow a middle course, avoiding extremes. As pointed out in the preceding chapter, we are organs of protest against everything crude and harsh that nature has produced, no less than organs of appreciation of everything excellent and lovable that it has brought forth. To exercise either of these functions to the exclusion of the other is to become lamentably one-sided. How we exercise them makes a great difference. To brood inertly over all the harshness and violence in nature and all the injustice and suffering among men is vacuous. At the other extreme, to plunge with great energy into some large philanthropic undertaking, thoughtless of the harm we might meanwhile be doing in other quarters or of future ills that might follow from present benefits, is shortsighted folly.

The safest course of those sincerely concerned to alleviate suffering is so to live that we do not avoidably harm any creature. This course, which may involve denying ourselves certain indulgences and pleasures that our neighbors do not consider wrong, is the most convincing protest against harshness and violence that we can make. Beyond this, if one is able to alleviate suffering or bring cheer, even on a small scale, such active endeavor should distract a sensitive spirit from futile brooding over all the ills that he is powerless to correct. A single kindly act is worth more than a whole philosophy of despair.

This course should leave us free to appreciate, without qualms of conscience, the fairest productions of nature and of man. Not by lamenting its failures but by celebrating its achievements do we show our loyalty to the Whole to which we belong. Appreciation is not a passive but an active state, demanding attention and understanding; when most sincere, it leads to the further activity of caring. I know of no more fitting way to show appreciation and gratitude than by helping others to understand, appreciate, and be grateful and by encouraging them to cherish and preserve whatever seems most worthy of being loved or admired.

A thoughtful person's happiness depends, above all, on fulfillment, the feeling that he is not wasting his life but enriching it with experiences and achievements of undoubted worth. In youth, this feeling is promoted by growth, learning, and developing skills; in our middle years, by progressing steadily toward whatever goals we set for ourselves; in old age, by remembering solid accomplishments and gratifying experiences and being surrounded by those we have helped to live well. This sense of fulfillment contributes more to a serious person's happiness than many transitory pleasures and may be even more important than health and soundness of limbs. When sick or injured, we may be more distressed by our inability to advance our undertakings than by physical suffering. Despite bodily defects or chronic poor health, many people have managed to live contentedly because they were able to achieve things that mattered to them. Epicurus, dying of a painful illness, declared that his last day was his happiest. Surrounded by devoted disciples, he could look back upon a highly successful career as the teacher of a discipline for living tranquilly, without fear of death.

A philosophy more profound than that of Epicurus, who regarded all composite bodies as outcomes of fortuitous encounters of atoms, is needed to make us adequately appreciative of a happy, rewarding life. To suppose that organisms as complex as ourselves, with such great ability to perceive, enjoy, and cherish, arose by chance in a world that offers so much to sustain and delight us is shallow thinking. To be sure, random events, such as genic mutations, have played a large role in evolution, but, unless a persistent tendency to raise existence to higher levels of organization and awareness had seized upon each mutation that could advance life, we and most of the things that make our lives rewarding would not be here.

The most urgent present task confronting those who care is to achieve a balance between appreciative minds and things worthy to be appreciated. The more units in either of these categories that our planet supports, the greater the fulfillment of the creative process. Unfortunately, too many of the first is certain to reduce the abundance of the second. There is, happily, a growing tendency to find joy and fulfillment in contacts with the beauty, sublimity, wonder, and tranquility of unspoiled nature. To support the kind of culture in which appreciation and understanding of nature most flourish, we must alter the natural world on a large scale, by planting crops, supplying timber for construction and pulpwood for paper, and in many other ways.

Up to a certain point, such alterations need not be disastrous; they may even augment the number of things to be admired and appreciated. A landscape pleasantly diversified, with cultivation in the valleys and woodlands on the hills, can be more delightful than a vast extent of unbroken forest, in the midst of which one enjoys no outlook at all. A varied terrain supports a greater variety of organisms than one more uniform, for many plants and animals that cannot endure deep shade thrive in the clearings, and not a few kinds of birds emerge from the woodlands to sing and nest in fruit trees and gardens. Ease of travel enables us to experience more widely and appreciate more fully the wonderful diversity of nature and the most impressive achievements of man.

As populations increase, the ideal balance between cultivation and wild nature is upset. Woodlands are destroyed even on soils unfit for sustained agriculture, in the vain attempt to produce enough food for man's hungry generations. The phenomenon is not new—two thousand years ago, Lucretius, in his poem *On the Nature of Things*, remarked how yearly the woods retreated up the mountainsides—but in recent decades the alarmingly rapid growth of populations has caused deforestation at an unprecedented rate. Even if only a small minority of people seek contacts with nature, the number is now so great that parks and reservations set apart for their recreation suffer from too many visitors. Birds are endangered by the growing multitude of people who wish only to watch and admire them. As the balance is upset, more and more people are reduced to enjoying less and less.

The most tragic aspect of the present situation is that, as populations swell and great masses of people are forced to live in unwholesome concentrations, with little to sweeten an environment horrible with noise, ugliness, pollution, squalor, and unrest, the capacity to appreciate and to care is supplanted by hatred, violence, and destructiveness. It is obvious that, after a certain critical point has been passed, the number of appreciative, cherishing minds cannot increase in proportion to the total population. Humanity must inevitably deteriorate along with the natural world that supports it.

Since appreciative, cherishing minds appear to give its highest significance to the aeonian movement that produced them and to be its justification and fulfillment, they cannot become too numerous, so long as they can exist with no undue diminution of the things worthy of their appreciation and concern. To maximize both these categories, we must take a long view. The visible alternatives are huge masses of hungry, restless people, in a world that fosters violence rather than appreciation as it plunges toward global disaster, or a moderate density of well-bred, well-educated, adequately nourished people, who make upon nature's productivity restrained demands consistent with the preservation of its beauty and diversity. By the latter course, the number of appreciative minds that Earth can support over the ages can be immeasurably increased. Apart from our immediate dependents, we have no greater responsibility for our contemporaries than for our successors. Those who care greatly because they appreciate greatly have no more sacred obligation than to do everything in their power to preserve the kind of world that will nourish appreciative minds for countless generations.

With infinite travail, the world process made us able not only to enjoy and appreciate but likewise to advance with foreseeing minds its agelong striving to make existence precious. By exercising these abilities, we give high significance to our own lives while we increase the value of the Whole to which we belong. For the privilege of living in a world created with such immense effort and participating even slightly in its creative advance, we should be ineffably grateful. Appreciative, cherishing minds are the world's best hope.

In chapter 2, I maintained that Earth's importance springs from the fact that it is the chief expression point, in a vast expanse of space, of what the creative energy can accomplish when it finds favorable conditions. Now I shall close this book by asserting that man is to Earth what Earth is to the solar system, the major expression of what evolution has so far achieved on this planet—at his best, of what is fine and admirable; at his worst, of what is diabolically evil. Although it is easy to exaggerate the cosmic importance of man, it is equally easy to underestimate it. Considered merely as an animal that is born, eats, grows, reproduces, uses tools to make artifacts more or less elaborate, and dies, man is just an extremely interesting, but often far from lovable, addition to the vast array of animals that inhabit

Earth, no more important than any of the others but, unfortunately, with a much greater capacity for destroying the others and polluting the environment. Nevertheless, viewed as a spiritual being with great powers of appreciation and an unquenchable determination to find significance in his own existence and that of the wider Universe, man, at his best, is a unique being, of truly cosmic importance. For the Universe would be much poorer without minds that gratefully celebrate its beauty and sublimity, are concerned for the preservation of the minute fraction of it that is accessible to them, and stubbornly try to understand it. And where, apart from man, do we find such minds?

Index